Reconstructing Environmental Governance

Reconstructing Environmental Governance

The Chance to Choose a Better Future

Rick Reibstein

ROWMAN & LITTLEFIELD
Lanham • Boulder • New York • London

Published by Rowman & Littlefield
An imprint of The Rowman & Littlefield Publishing Group, Inc.
4501 Forbes Boulevard, Suite 200, Lanham, Maryland 20706
www.rowman.com

86-90 Paul Street, London EC2A 4NE

British Library Cataloguing in Publication Information Available

Library of Congress Cataloging-in-Publication Data

Names: Reibstein, Rick, 1952- author.
Title: Reconstructing environmental governance : the chance to choose a better future / Rick Reibstein.
Description: Lanham : Rowman & Littlefield, [2022] | Summary: "This book takes a creative approach to policymaking, urging citizens to engage fully with democratic governance to create more effective environmental protection programs. The technical and economic feasibility of transforming our practices to develop more sustainable and nurturing societies encourages us to jointly envision a better future"— Provided by publisher.
Identifiers: LCCN 2022014112 (print) | LCCN 2022014113 (ebook) | ISBN 9781538160039 (cloth) | ISBN 9781538160046 (paperback) | ISBN 9781538160053 (epub)
Subjects: LCSH: Environmental policy. | Environmental protection—Citizen participation. | Environmental protection—Political aspects.
Classification: LCC GE170 .R4452 2022 (print) | LCC GE170 (ebook) | DDC 363.7/0525—dc23/eng20220820
LC record available at https://lccn.loc.gov/2022014112
LC ebook record available at https://lccn.loc.gov/2022014113

Contents

Preface

INTRODUCTORY COMMENTS

This is a book about environmental policy. It has a lot of references to things that environmental policy professionals might want to know, but I hope anyone who might be interested in having better environmental governance might pick it up. It has a lot of personal stories as well as history and some of the political drama of what we think of as regulations but are much more than that—governance is whatever government does. All citizens in a democracy are part of that, whether they realize it or not, and all members of the human race are part of the environmental impact on the planet, and can be part of the solution.

When I decided in 1970 that being an environmentalist was what I wanted to do, people responded favorably (except my parents, worried I would starve). People were supportive of environmentalism and that positive response has been dominant throughout the decades of my career. But there is another attitude that is not often expressed to me outwardly, but sometimes suggested, often as a joke: environmentalism is irritating and, possibly, too pessimistic. It's costly. We have to keep it under control. This attitude, reflected in our politics, has had real impact and led to budget cuts and constant suits. The shared mission of environmental protection has been badly degraded and is in need of protection. The bigger view that reminds us we are all in this together and in some very real degree of trouble requires restoration.

But it needs more than just restoration. We should re-engage in an active effort to evolve ever more effective approaches. This book is about that idea. It covers general concepts related to improving environmental governance and offers examples from a career in that effort. My story is part of a larger story that people should know. Biden has opened the door to remaking our system, but Clinton and Gore did that in a big way decades ago, and so have many others. We should draw on decades of thinking and experience. One example, prominently featured in this book because of my personal experience, is the nearly forgotten but incredibly successful pollution prevention movement. But there are many more, involving better ways of solving problems and making decisions as a community. We should be engaged in a national review of what has worked, so that what we build now is informed by it.

My hope is that these stories, references, and facts help foster interest in continuously improving our system of democratic self-governance, which is the key to environmental salvation. A focus of intellectual curiosity on what ideas and innovations have been successful is what we need, for without that the risk is high that we will re-create inefficient modes and slow the necessary transition to sustainability.

For those who want to know the main argument, and those who like their points in order:

1. Despite the problems of misusing government, there are also government programs that do great things.
2. This is our democratic inheritance as a people: the chance to have good government programs.
3. We must not give that up because it hinders some business plans and we must not fail to try because so many government programs have not succeeded. We have to try to get better at it.

At the outset, the issue of this book's attitude toward business must come up in many minds, as too many tend to frame the discussion as "jobs v. environment." Business planning deserves great respect, but when the question is what comes first—the opportunity to do business or the fundamental needs of life itself, or the rights of people—we can recognize how *commerce can adjust* to properly set priorities, but life cannot adjust to seriously warped priorities. Ecosystems are not adjusting well to the improper priorities of continued extraction and emission, consumption and waste generation of the "throughput" economy.

Visions of the circular economy, of the "nested" or "doughnut" economy, of ecological and humane and responsible economies, now abound. They are not just theoretical but may seem that way to many people. They reflect a real evolution that can peter out or accelerate, depending on whether enough people pay attention, care, and act. If we can actively envision implementing the better future, constructed from the context of a more humane and sensible economy, we increase our chances of attaining it.

I was lucky to be hired in 1988 in what became the Massachusetts Toxics Use Reduction (TURA) program, which worked with thousands of businesses and others to develop less toxic ways of doing virtually everything we looked at. It worked.[1] The reduction in billions in pounds of toxics use by such programs across the country would not have happened without an investment in innovative governance. The world of clean energy, safe products, intelligent housing and transportation, humane and sensible food systems, replenished wildlife, beauty, and stable climate is achievable through the help provided by the implementation of relevant government services in a framework of well-designed rules. But while some pray that tech wizards and business tycoons will solve our problems for us and I observe with wonder what they do sometimes produce, I believe they can be a force for good but may not be unless we revalue democratic government itself and use it to establish programs to nurture the development of responsible activity.

The growth of innovative environmental programs peaked in the 1990s, diminishing because of budget cuts. That flowering of creativity and understanding, of mutual respect and problem-solving, of consensus-seeking, can be revived, for it still exists.[2]

The stated purpose of our democratic republic was to serve the general welfare. The joint loss of natural life and its stability is something we must address, with our government. Because of the gift of the founders, we can.

Partisanship for the Earth

This is not a nonpartisan, nonpolitical, neutral, objective text on environmental issues. It is written for those who see the need for government to act so that businesses, governments, organizations, and people—we—stop harming the ecosystems we need to survive. The question is not whether to act but how best to accomplish this aim—with the maximum freedom, the least tax burden, the most efficient programs, the most just results, and the widest benefit. This book has that point of view. It is avowedly partisan for the Earth, and for the idea that in order to protect the Earth we must transcend political partisanship and strongly manifest our common interest in protecting our shared environment.

1. See program assessments, progress reports, and so forth, at https://www.turi.org/TURI_Publications?Topic=TURA+program and/or annual data reports at https://www.mass.gov/guides/massdep-toxics-use-reduction-program.

2. As an example, I was on the board of the National Pollution Prevention Roundtable during the 1990s when it received adequate funding to have several national and regional conferences and launch initiatives. When I rejoined the board seven years ago, it was in the red, and we thought it might go out of existence. Now it has revived and, though not well financed, has become active again. See www.p2.org. Read stories of Most Valuable P2 awards. The flame has not burned out.

This is urgent because we all face a shared environmental emergency. We have inherited processes of democracy, civil discourse, reasoning, and seeking consensus and using them is how we find the right balance of freedoms and necessary restraints that can gain widespread willing acceptance. This is what we should seek in the belief, held by the founders of democracies and the philosophers who inspired them, that the legitimate government is that which governs by consent of the people. This vision has to be articulated because we are currently suffering from a politics that is about contest, winning by any means. Efforts to suppress voting, for example, or to refuse to accept the proven results of an election, reverse the trend to manifest this principle, and seek to return us to a more primitive, illegitimate form of government. Because the environmental emergency faces us all as one body of humanity it is imperative not to fragment now but to transcend divisions and unite around our common interest in surviving without worsening weather, famine, poisoning, oceans of waste, or loss of nature.

This approach makes membership in the Democratic, Republican, or any other party less relevant than membership in the human race or in a body of citizens. The current opposition of the Republican Party to environmentalism and to democratic fundamentals (such as voting, acceptance of court decisions, findings of facts, science, etc.) creates a need to articulate our actual, vital context.

It is not faction, as the founders worried, that destroys, but *excessive* faction. That is the meaning of the "loyal opposition." You argue your point and then bow to the will of the majority vote, and you only keep fighting when the majority wrongly infringes on fundamental rights of the minority. Fighting forever, over everything, so that your party wins no matter what principles are lost, never giving in, never compromising, never affording your opponent a "win," is un-American, if America is to consider itself a democratic polity or a republic, in which representatives of the people work for the interests of the people.

Partisanship for democracy itself has to be articulated and asserted, and so does the principle of Earth Partisanship, because the public discussion tends too often to quickly devolve into a binary choice between two positions. The perspective of listening to all sides has to be articulated in order to create a transcendent context in which agreement can be found, and without this framework too many people descend into a battle focused on producing only one voice and extinguishing the other. Our governance could be more creative, it could generate more shared understanding. It could do more to bring us together in public meetings and facilitate cooperative efforts. These all take investments in staff, but it's needed. Real relationships need to be created and facilitated.

The old song asks, "Will the Circle Be Unbroken?" The circle of life *has* been broken. We must restore it. This is a mandate of reality. Our entire society should, and can, take on this task as a common duty (differentiated to some extent according to context). That can't happen if we are fighting each other. Imagine how much progress could take place if scorched-earth politics were rejected, if manipulations of anger and fear were recognized and deflated.

The Competitive Context

Another thing to address at the outset is the concept of competition. The idea that we should exist within a competitive context competes with the fact that we coexist. We need to get the context right. The environmental emergency requires the cooperative context be recognized as primary. Continued fragmentation from exclusive allegiance to a sub-universal context prevents our developing mature societies capable of responding in an adult manner to the problems we must face together.

Attention only to the competitive context leads to arguments for government nonintervention in business affairs. History shows that competition subsumed within cooperation is beneficial—imagine a well-refereed game. But *unrestrained competition* can be harmful—imagine violence on the field. It is a fact that there are irresponsible political actors who have disseminated disinformation and ideas harmful to public understanding of the value of continuing investment in environmental law.[3] They are pursuing the logic of unrestrained competition, as the tobacco and lead companies did when they disputed the science about the harm of their products, going so far in their fervor as to attack honorable scientists just doing their job.[4]

No reader of this book should think it is an antibusiness tract. It is opposed to the *religion* of business, to the *primacy* of money, to *unrestrained* capitalism, not to business or to money. I met many business leaders, as an environmentalist, in my career, and have great respect for a great many of them. But they all had to put profit first and needed governance, and generally recognized that. Asserting the primacy of people and ecosystem health does not extinguish economic goals, but puts them in their proper place.

---

3. See, for example, the Union of Concerned Scientists' Climate Deception Dossier, https://ucsusa.org/resources/climate-deception-dossiers.

4. Such as Dr. Herbert Needleman, who showed us how lead was harming the brain, was harassed and accused: Carey, Benedict. "Dr. Herbert Needleman, Who Saw Lead's Wider Harm to Children, Dies at 89," *New York Times*, July 27, 2017, https://www.nytimes.com/2017/07/27/science/herbert-needleman-dead-lead-poisoning-in-children.html; generally, see Stephen Burd, "Scientists See Big Business on the Offensive," *The Chronicle of Higher Education*, December 14, 1994, https://www.chronicle.com/article/scientists-see-big-business-on-the-offensive/.

A Constitutional Perspective

The founding of this republic, as flawed as we see it today, yet revealed a deep and potentially saving truth: that when aspirations of different interests are joined, we approach universality. The Preamble to our Constitution reads:

> We the People of the United States, in Order to form a more perfect Union, establish Justice, insure domestic Tranquility, provide for the common defence, promote the general Welfare, and secure the Blessings of Liberty to ourselves and our Posterity, do ordain and establish this Constitution for the United States of America.

It does not say, We, the property holders of this country, now institute a system of laws that will protect our property and preserve our authority, as is the apparent purpose of government in many countries today. It expresses, instead, a set of universal values, and locates the legitimacy of governance in the people as a whole.

In a class I teach called Law for Sustainability, I ask my students, What is the most accurate way to describe the legal effect of the Preamble to the Constitution?

a. It is a guarantee that can be used in support of a claim of environmental rights.
b. It supports arguments for limitations on the powers of the federal government.
c. It is a guide to interpreting the meaning of the document.
d. It has no legal effect because it is not "self-enforcing."

The meaning of the Constitution is in the words that the framers used to introduce it. Nevertheless, the courts have not used the Preamble to give direct legal effect to its stated aims. It has neither acted as a guarantee nor limit, and some commentators have actually said it has no legal effect.[5] But courts have seen meaning in it, so that "c" is the best answer. As Erwin Chemerinsky, dean of Berkeley Law School, and Michael Stokes Paulsen, of the University of St. Thomas School of Law, write on the Constitution Center's website:

---

5. For example: "As the Founders intended, the Preamble has no force in law. It grants no powers to the federal or state governments, nor does it limit the scope of future government actions. As a result, the Preamble has never been cited by any federal court, including the US Supreme Court, in deciding cases dealing with constitutional issues" (Robert Longley, "Preamble to the US Consitution," *ThoughtCo.*, last updated September 4, 2020, https://www.thoughtco.com/preamble-to-the-us-constitution-3322393.)

> Importantly, the Preamble declares *who* is enacting this Constitution—the *people* of "the United States." The document is the collective enactment of all US citizens. The Constitution is "owned" (so to speak) by the people, not by the government or any branch thereof. *We the People* are the stewards of the US Constitution and remain ultimately responsible for its continued existence and its faithful interpretation.[6]

This principle has been recognized since the earliest decisions of the Supreme Court. The court has since then generally used the Preamble to confirm and support its interpretations of provisions of the Constitution. This can be done conservatively, staying with the terms that are explicit, and one can argue that has been the general practice. But it can also be used to support that which we should regard as *necessarily implied* by the Constitution's themes.

I also ask students to look for environmental protection in the Constitution, and most of them find it, not explicitly stated but inherent in meaning, present by necessary implication. For example, the term "General Welfare" is not just in the Preamble but in the powers given to Congress. Students ask, *How can we protect the General Welfare without environmental protection?* The same reasoning goes for all the other powers of Congress, which all flow from the purposes stated in the Preamble, and if we cannot muster the vision to perceive what is necessary in order to achieve them, we discard the great gift of the founders, preserved and summarized by Abraham Lincoln, as a government of, by, and for the whole people.

The tools of governance include: regulation, grants, cooperative research and development, convening meetings to produce consensuses and road maps, financing, assisting, establishing preferences and tax breaks and loan guarantees and bonds and training, educating, combating misinformation, facilitating forums, rewarding innovative design, demonstrating and recognizing successes, certifying desirable actions and verifying claims, requiring reporting, organizing information, monitoring, analyzing, starting public-private ventures, forming cooperatives, preventing poisons from entering our lives, preserving nature, and preserving health, as well as imposing limits on behavior.

In the mid-1990s, when I was working in the Massachusetts Executive Office of Environmental Affairs, we were visited by representatives of a trade association for companies that make household and consumer products. They told us that they were concerned about our Household Hazardous Waste program, which advised people not to buy products that become hazardous waste. They suggested they might sue if we couldn't back up claims we made. So I applied to the Environmental Protection Agency (EPA) for a

6. "The Preamble," Interactive Constitution, National Constitution Center, https://constitutioncenter.org/interactive-constitution/interpretation/preamble-ic/interps/37.

grant to create an effort to identify environmentally preferable products, so that we would have what we called "defensible" reasons for making recommendations on what people should purchase. The initiative became part of a new effort we had just launched to improve the environmental performance of state agencies. It is now a large part of the states' purchasing arm.[7]

It didn't solve the problem of household hazardous waste: people still need to know that states have identified environmentally preferable products for agency use, and they can also replace hazardous products with safer ones. We need to reestablish vigorous household hazardous waste programs throughout the country and get that terrible waste stream down. *The point is that the EPP programs were a new tool of governance, and they have helped foster the growth of a green economy.* Companies that were making cleaner and safer products were given a market, and fair methods for qualifying as preferable. The superior environmental performance that innovative companies come up with gets rewarded.

Government has many tools it can use, not just the familiar ones of setting standards and punishing people for not meeting them, the ones that make environmental governance seem like a restriction on enjoying life, that make the EPA like the enemy. We can make new tools, find new ways to foster, to nurture, to cultivate, what will help transform our world into a network of sustainable societies. It is necessarily implied by the Preamble that we need to use government creatively if we are to protect our general welfare.

As we move to restore the governance that corrupt administrations tear down, we can build them back so they are relational, preventive, and efficiently effective. We have the right as a people to act to foster and favor the growth of the clean and circular economies we need in order to stop fouling this Earth and to treat it like the miraculous gift that it is. Contemplating the failure to climb out of our dirty past is necessary but does not condemn us to remaining in it. Anger, mourning, disappointment, resentment, impatience—which all threaten to drive us into depression—can yet be useful if, instead, we move on to discover the distinct pleasure of working together to create a future of better prospects.

---

7. The original EPA grant that paid for an intern (Eric Friedman, who went on to lead the Commonwealth's Leading by Example program) was in 1994. I credit attorney Jonathan Goldfield, then of the Operational Services Division, for immediately perceiving the value of the idea and giving the program a home in the purchasing office. In 2009, a Governor's Executive Order formally established the office and made use of proven environmentally preferable products (EPP) mandatory for state agencies. https://www.mass.gov/doc/2019-annual-report-for-ma-epp-program/download. The 2011 annual report notes that "Since the Program's inception in late 1994, thousands of EPPs have been incorporated into an estimated four dozen statewide contracts and the state's purchases of these goods and services have grown from $5 million to over $305 million per year." On EPP generally, see the National Association of State Purchasing Officers: https://www.naspo.org/green-purchasing-guide/ or the Responsible Purchasing Network.

A Note Concerning Efficiency and Morality

It is not unusual in a class that covers environmental economics to introduce students to the concept of Pareto optimality, also known as Pareto efficiency, named after the Italian economist Vilfredo Pareto. The *Encyclopedia Britannica* tells us:

> He laid the foundation of modern welfare economics with his concept of the so-called Pareto Optimum, stating that the optimum allocation of the resources of a society is not attained so long as it is possible to make at least one individual better off in his own estimation while keeping others as well off as before in their own estimation.[8]

It's a really interesting idea. But there are other ways of thinking about optimality, such as through a direct focus on whether a society is just.[9] John Rawls was an American philosopher who produced *A Theory of Justice* in 1971. The *Stanford Encyclopedia of Philosophy* describes his contribution:

> His theory of *justice as fairness* describes a society of free citizens holding equal basic rights and cooperating within an egalitarian economic system. His theory of *political liberalism* explores the legitimate use of political power in a democracy, and envisions how civic unity might endure despite the diversity of worldviews that free institutions allow.[10]

The duty of government is to be efficient, and the idea of Pareto optimality appears to be helpful in thinking through how that goal might be attained—at least in terms of social acceptance. The duty of government is also to be just, and Rawls's vision helps us see more deeply what that direct relationship with justice, with fairness, can be, and he provides simple concepts that anyone can use. One is the veil of ignorance—that we would design a better society if we don't know what position we will have in it. If we know we are born rich, we might find ourselves supporting an unequal society. If we don't know if we will be born rich or poor, we would likely choose a more equal one. Another concept is that we should all enjoy the fullest extent of liberty, until our freedom harms others. His concept of the Difference Principle is concretely illustrated by Nigel Warburton in *A Little History of Philosophy*:

---

8. *Britannica*, s.v. "Vilfredo Pareto," last modified August 15, 2021, https://www.britannica.com/biography/Vilfredo-Pareto.

9. For example, see Lawrence Sager, "Pareto Superiority, Consent and Justice," *Hofstra Law Review* 8, no. 4 (1980): article 3; Howard Change, "A Liberal Theory of Social Welfare: Fairness, Utility, and the Pareto Principle," *The Yale Law Journal* 110, no. 2 (2000): 173–235.

10. *Stanford Encyclopedia of Philosophy*, s.v. "John Rawls," last updated April 12, 2021, https://plato.stanford.edu/entries/rawls/.

> Society should be arranged to give more equal wealth and opportunity to the most disadvantaged. If people receive different amounts of money, then this inequality is only allowed if it directly helps the worst off. A banker can only get 10,000 times more than the lowest-paid worker if the lowest-paid worker benefits directly and receives an increased amount of money that he or she wouldn't have had if the banker was paid less. If Rawls were in charge, no one would earn huge bonuses unless the poorest got more money as a result. Rawls thinks this is the kind of world reasonable people would choose if they didn't know whether they would be rich or poor themselves.[11]

When we set out to improve the system by which we govern ourselves, we will have a greater chance of success in creating systems that are widely accepted, if they are *both efficient and just*. We can use the device of the veil of ignorance in the same way that two people can share a piece of cake fairly—one cuts the pieces and the other chooses. Democratic process can provide both efficiency and fairness if that is the shared intent of participants, and that is more likely to be forthcoming if public meetings are facilitated with these goals in mind.

PLAN OF THE BOOK

The plan of this book is to tell some stories about how government can work, so that more people will understand that we need to be engaged with government, supporting innovative government, so that we can have more programs that work, and so that we can all play a role in ensuring that they work well.

It is hoped that by focusing on the idea that we can all work together, through government, to address our shared problem of the environmental emergency, that we can all feel less lonely, less powerless, because we can find how capable we are when we cooperate.

The book introduces the idea that there's a story about good, effective environmental governance that needs to be told, and that there's a story about how environmental governance is bad, that needs to be seen as the falsity that it is. It is hoped that by discussing things the way they really are, how we have corporations that do behave unethically, and who support disinformation, and how we have good companies run by good people trying to do things the right way, readers will come to abandon cynicism and despair and will realize how we can structure a governance system so that it encourages the good and discourages the bad.

The book then urges the adoption of the big picture, using that largest perspective as our context, and it is hoped that it is made clear that this, done

11. Nigel Warburton, *Little History of Philosophy* (2011), 231.

right, in no way destroys or diminishes the individual or group identity. If we concentrate on government as nurturer, cultivating willing compliance, we maximize the flourishing of the individual, and this is good for the entire population.

A few more principles are discussed: that governance must be anticipatory, thinking about what can go wrong so we can avoid foreseeable harms; and government must be equitable, not just because it is right but because it is also sensible, providing stability as well as justice. We should think about the "quality of our awareness," which ought to be characterized by active envisioning of the future we could have, an accurate assessment of how we are seeing and understanding the world as it is, and building on programs that have worked.

The chapter called "The Fixer-Upper" is written with the idea that we may not be able to conjure up the political will for the wholesale reconstruction the rest of this book recommends, and essentially says, okay, if you don't want to do that work, at least let's make some fixes to long-standing issues. By no means are the examples in this chapter the only things that need fixing but they are issues with which I have had direct experience.

The last chapter is about how we need to transform all the systems we use for living, and some thoughts about how that can actually be done. It is hoped that readers will recognize that there's a whole literature out there, growing fast, about how we will end up better off if we go down the right road. I call it the Hump. The investment we need to change is a hill we have to climb. But on the other side is free coasting, in some cases, permanently, such as freedom from unnecessary poisoning, or saving species from extinction. All we have to do to see how worthwhile the investment would be is to imagine we are our own grandchildren.

Finally, there is an appendix that is just there to illustrate how many more issues and ideas there are to be considered once we start thinking about how to save this world from itself. There are also exercises at the end of some chapters, for those who teach or can use them at meetings.

ACKNOWLEDGMENTS

I am deeply grateful to my mother for making me an environmentalist (not her intention, I'm sure). She told me one day, when I was nine or so, to come inside if it rained because they were going to explode an atomic bomb and if it rained it could bring down radioactive fallout. When it started to rain, I told the other kids we had to go inside and they had no idea what I was talking about. I was the only one who left our game of soccer baseball to avoid the fallout. She also worked for the (city) government helping to plan, form policy, and develop programs, as I ended up doing, and took me to "non-partisan" debates held by the League of Women Voters, where I first saw and admired democracy in action. I am also deeply grateful to my father for insisting I choose a career that would not only be a way of earning a living but also something fulfilling.

My friend Erik Butler gave me a reference that got me my first job in government, though it was not something that I thought made any sense at the time—in the budget office? As it turned out, examining every environmental budget request from every agency was a fantastic start. It led me to the Office of Safe Waste Management, and I am grateful to Department of Environmental Management Commissioner Jim Gutensohn and Office Director Michael Brown for hiring me and giving me my start in "source reduction," which became pollution prevention. I am grateful to Barbara Kelley for being the best boss ever during what I call the Golden Age, when we grew from just the two of us to more than thirty, won awards, made many friends among businesspeople who wanted to do the right thing, and spread the good word. I want to thank George Frantz, particularly, for his great work and friendship during that time, as well as the entire population of the pollution prevention field, represented by the National Pollution Prevention Roundtable; but I will mention Ed Kunce and Nancy Wrenn of the Massachusetts Department of Environmental Protection, pioneers in prevention and better regulation; and Ken Geiser, who did so much to found the great Massachusetts Toxics Use Reduction program, and Liz Harriman and Suzie Peck. There were so many others who made it work. I mention as inspirations Nick Ashford, John Ehrenfeld, and Jennifer Nash, when they were all at MIT, and pioneering pollution prevention consultants like Robert Pojasek and Robert Capaccio, in an inadequate effort to indicate how many more people were part of the re-envisioning of environmental protection that must be recovered and reinvigorated. It is terrible to leave out so many other names that deserve mention, like Terri Goldberg at NEWMOA; or organizations like Canada's Pollution Probe, Joanna Underwood's INFORM, or the Institute for Local Self-Reliance. The history of pollution prevention needs to be written.

I also want to thank the other best boss ever, Deborah Brown, with whom I fought lead poisoning while at the EPA. I want to take this opportunity to thank Cutler Cleveland for giving me a lectureship at Boston University, and all the students who have made that position a positive joy. I thank my children for being interested in and respectful of my obsessions, supportive of my hopes, and loving. I thank my wife for the same reasons, but also for her unfailing and life-giving support. I did not share drafts of this book with colleagues, so there is no one else to blame for its shortcomings. But I have learned from so many. As I think about whether there are any so key to my thinking that I should mention them, I think about the thousands of government reports and memos I have read, or worked on, and how they were the product of teams, often with the primary authors buried in a note. I think about how my own contributions were often under the name of the official in charge, not my own. This is often the nature of government work, which has to be its own reward. That said, there could be more efforts to give credit where it's due. That might better incentivize innovation from the ranks.

Chapter One

The Facts Have Spoken

We Can Do It

The facts have spoken, and they have said that we can we have a sustainable world. The economics of renewable energy are such that wind and solar are now cheaper than fossil fuels.[1] We don't need a "bridge" to renewables made out of something poisonous, such as more radioactive-waste-producing nuclear plants or geologically mined gas. We can go right to clean energy. Why we don't is the matter before us, not whether we can.

We benefit by recognizing human nature. We can see that there's a lot already invested in the older methods and it is hard to change. But some of the resistance to change is unethical—such as funding climate change denial[2]—and some is specious, such as the argument that renewables are not cheaper without government support. This story works with many who do not know that far larger and steadier subsidies have gone to conventional energy.[3]

---

1. "The World is Adopting Renewables Faster than Experts thought Possible," subheadline for "The Wind and Solar Boom is Already Here," by Farhad Manjoo, *New York Times* (Op-Ed), April 30, 2021. "More than half of the renewable capacity added in 2019 achieved lower electricity costs than new coal, while new solar and wind projects are also undercutting the cheapest and least sustainable of existing coal-fired plants. Auction results show these favorable cost trends accelerating, reinforcing the case to phase-out coal entirely." *Renewable Power Generation Costs in 2019*, International Renewable Energy Agency, https://www.irena.org/-/media/Files/IRENA/Agency/Publication/2020/Jun/IRENA_Power_Generation_Costs_2019.pdf.

2. For example, see Douglas Fisher, "Dark Money" Funds Climate Change Denial Effort," *Scientific American*, December 23 2013, on how the money is funneled so that its source is not easily determined, https://www.scientificamerican.com/article/dark-money-funds-climate-change-denial-effort/.

3. Such as, the oil depletion allowance, or the supply of nuclear materials from government operations, or the exemptions from environmental rules enjoyed by the fossil fuel industries (or lax treatment thereunder, such as allowing mountaintop removal), or limiting of liability for nuclear accidents and oil spills. The list goes on. Wal Van Lierop, *Forbes* contributor, writes "Most people are unaware that their tax dollars subsidize fossil fuels," https://www.forbes.com/sites/walvanlierop/2019/12/06/yes-fossil-fuel-subsidies-are-real-destructive-and-protected-by-lobbying/?sh=5177490a417e.

How can we place our all-too-human desires in the context of biospheric realities? We can restore primacy to the needs of humanity—of living ecosystems—without falling into the trap of debating a binary choice of being purely capitalist or communist. The desire for simplicity, the habits of identity and ideology, these and other social phenomena (such as some forms of mass communication) can reduce our thinking to this absurd choice. The truth is, we have many choices to consider—so many ways we can balance our interests with the need to stop environmental damage. A first step is to recognize and call out the absurdity of a reductive argument, such as that any governmental action to control business is the "slippery slope" toward communism. A great storehouse of ideas awaits our sophisticated analysis so that we can choose which aspects and methods of governance to combine for best results.

Instead of seeing politics as a battle between who wins and who loses, we can think of each problem as a struggle for the population to mature as a species, which will not be through the pitting of parties against one other, as in trial by combat, jury, vote, or fashion, but by the optimal satisfaction of everyone's needs. Adversarial processes have their place, but a universal movement to transition to a fair and green world, which will only be created through cooperation, is not yet forceful enough to counter the shortsighted. We have to generate the spirit of it. This need not be thought of as impossible. For one thing, the idea of working together is appealing and especially when contrasted with conflict. It can be fun. For another, it makes sense to welcome all people on the journey, for it is all the people evolving into a sensible humanity that the living Earth needs.

There is also nothing about the need for change that necessitates that it be sudden, or rough, or conducted as warfare, with enemies. Though it requires defense at times, it is the development of a new art, of creative change, of gentle transformation, that can be pursued with some expectation of success. Enough of this approach has worked that we can realistically set as a goal curtailing the activities of the powerful fossil fuel companies, mining companies, and others who extract resources or value or otherwise damage our common heritage. The groups of stakeholders in the fate of the planet—communities, parents, states, tribes, voluntary organizations, schools, cooperatives, farmers, workers, citizens, and so on—can, united, create new rules for economic behavior, because that is their right and duty.

Abandoning binary and adopting multiple perspectives, tolerating complexity, will allow the more easily grasped universal principles to be widely seen and used. One such principle is that development, comfort, sustenance—all these things we want and that we get right now from the modern economy—need not disappear if we place limits on destructive behaviors. A related principle is that these desired qualities *will disappear if we don't* place limits on behavior. A third important principle that must go with these is that we can design limits that maximize human freedom within the constraints demanded by morality and biophysical fact.

A Great Transition[4] that can address our problems can evolve gently and carefully. Seeing this helps in the gaining and maintaining of the will to move forward.[5]

We can also stem the great loss of biodiversity, particularly if we get moving on curbing greenhouse gas emissions. For example,

> Successful rebuilding of depleted fish populations has been achieved at local and regional scales through well-proven management actions, including catch and effort restrictions, closed areas, regulation of fishing capacity and gear, catch shares and co-management arrangements.[6]

Much work needs to be done—for example, tiger protections are failing, due to low investment.[7] But increasing the effort, such as by addressing poaching, the biggest threat (to many species, not just tigers), will make a difference.[8] When nature is protected, it often comes back.[9] We also see that where Indigenous Peoples have maintained their traditional presence on the land, biodiversity is also maintained, and when Indigenous Peoples are engaged in development, sustainable harvesting is generally what they choose to do.[10]

We can protect much of the precious biodiversity that remains by protecting the right of the people who live in those lands to remain there and help them develop the means, as these are cultures showing a continued dedication and respect for the land, and they give us the examples from which we can learn. This, of course, reverses the old, perhaps unexamined myth that

---

4. See, for example, Tellus Institute's Great Transition Initiative, https://www.tellus.org/.

5. Contrast this vision with angry demands, raised fists, and whatever people think of when they say they've had enough and they're going to "take to the streets." I wish I had a dollar for every time someone said that who I know would never wish to engage in or experience violence of any kind.

6. Duarte et al., "Rebuilding Marine Life," *Nature* 580, April 1, 2020, https://www.nature.com/articles/s41586-020-2146-7.pdf?origin=ppub.

7. https://www.worldwildlife.org/press-releases/new-study-shows-over-a-third-of-protected-areas-surveyed-are-severely-at-risk-of-losing-tigers.

8. Guardians of animals need protection, such as insurance and the tools to work safely and effectively.

9. See, for example, the Campaign for Nature—https://www.campaignfornature.org/why-30-1—and particularly its *Local Biodiversity Outlooks* stories, which present the perspectives and experiences of Indigenous Peoples and local communities.

10. Somini Sengupta, Catrin Einhorn and Manuela Andreoni, "A Global Plan to Conserve Nature is Lacking a Vital Ingredient," *New York Times*, March 12, 2021. The vital ingredient is inclusion of Indigenous Peoples in a meeting of dozens of countries to plan conserving a third of the planet's land and water. Famed biologist E. O. Wilson recommended a 50 percent goal to save most of the world's species (see https://www.half-earthproject.org/). See https://www.culturalsurvival.org/ for an example of an organization helping Indigenous Peoples to develop as they wish.

the modern world is more advanced. It also reverses the mindset of green colonialism (or eco-imperialism), in which people are forced off their land in the name of conservation.[11]

We can also change over to clean production with green chemistry,[12] toxics use reduction, extended producer responsibility, ingredient right-to-know, and policies that foster the evolution of a circular economy. These points are evident to anyone who has read the various relevant information—but you have to look for it. They are not prominently before the public in the national discussion. The right wing contemptuously dismisses these ideas as dangerous fantasy and skeptics abound also on the left, who worry that any idea that sounds too good will be misleading.

I probably would not have had the attitude that fixing our bad situation is entirely feasible if I hadn't had the luck of having a job in government providing assistance to companies to prevent pollution. This, which seemed an impossible task at first, turned out to be doable, and it showed me thousands of examples of how change is possible. A vast literature of pollution prevention (or, waste minimization, clean production) case studies exists, though much of it is not readily findable anymore.

Decades of telling others about what I have seen has shown me many ways that people block the information that beneficial change is feasible, available, or doable. It is as if people think that if we invest in our hopes, we will be disappointed, so it's best not to. But that's, ironically, like believing what Voltaire's Pangloss declared—that *this* is the Best of All Possible Worlds—and that's the attitude we're laughing at when we watch or read *Candide*.

There is a joke about an economist walking past a twenty-dollar bill and not picking it up, and upon being asked why he didn't, he said it couldn't have been there, because someone would have already picked it up. A great many companies I visited were actually leaving twenty-dollar bills all over the floor. It wasn't because they were stupid, it was because they were too busy chasing hundred-dollar bills. There was a rationality to it. But the point is, anyone who thinks companies do everything they ought to do because money is a force for efficiency, in all respects, is wrong. We know already that, because of costs the companies don't have to pay, such as wastes or pollution that harm others for which they are not held accountable

11. For example, https://foreignpolicy.com/2018/07/16/how-conservation-became-colonialism-environment-indigenous-people-ecuador-mining/, or https://www.researchgate.net/publication/280922007_Conservation_as_Economic_Imperialism. On how it can be done see https://news.climate.columbia.edu/2014/01/02/why-conservation-is-not-condescension-a-case-against-eco-imperialism/.

12. See, for example, EPA's 2020 Green Chemistry Award winners: https://www.epa.gov/newsreleases/epa-announces-winners-2020-green-chemistry-challenge-awards.

("externalities"), there is another concept of efficiency—for society and the environment. But in addition, they are also prone to inefficiencies relevant to the task of making money because they don't have the time to focus on what seems like small stuff compared to getting that product out the door. Companies need help seeing the small stuff, and sometimes that leads to big stuff.[13]

We can also attend to the broken cycles of nutrients, such as nitrogen and phosphorous, to stop overloading waters and depleting soils.[14] There are things we can do.[15] We can also transition to cleaner transportation and shelter, and curtail overconsumption, and address serious inequities in resource allocation, use, and impact. We can also ban plastics that are not totally recovered and reused, and devote resources to removing, as best we can, the plastics catastrophically littering this entire planet.[16] These tasks are all hard, but enough has been done to show that addressing them is doable and we have successes to build on.

We often settle for weak compromises that only partially address a problem. In 2011, Congress lowered the amount of lead in plumbing pipes and fixtures from 8 percent to 0.25 percent.[17] The industry was given till 2014 to clear their shelves of existing inventory before the restriction came into effect. This avoided economic havoc but allowed for the continued sale of equipment found dangerous for people to use. Congress could have appropriated funds to compensate the industry. It could have made the ban immediate. Instead, some continued poisoning was allowed when there had been a societal decision that there should not be that kind of thing happening.

---

13. For example, the textile firm we helped reduce toxics, which landed a contract with Patagonia as a result.

14. Conserving soil is related to both carbon sequestration and food policy. As Susan Schneider, founder of the Food Law Professors' LISTSERV wrote to colleagues in April 26, 2021: "I believe that we are at a moment in time when forces (good and bad) have come together to provide us with a unique opportunity to re-examine our food security and nutrition policies. Please note the following hearing that is coming up on Wednesday, April 28." (Rules Committee Chairman James P. McGovern (D-MA) announced a hearing on *Ending Hunger in America: Challenges, Opportunities, and Building the Political Will to Succeed*).

15. For example, the recommendations of the US Golf Association: https://www.usga.org/course-care/turfgrass-and-environmental-research/research-updates/seven-things-you-can-do-to-reduce-nutrient-loss.html.

16. For an overview of one aspect of this problem, see Katie Okamoto, "Your Laundry Sheds Harmful Microfibers. Here's What You Can Do About It." *New York Times Wirecutter* (blog), updated August 5, 2021. https://www.nytimes.com/wirecutter/blog/reduce-laundry-microfiber-pollution/.

17. The Reduction of Lead in Drinking Water Act amended Section 1417 of the Safe Drinking Water Act.

ROOT CAUSES AND NO-REGRETS STRATEGIES

We can address the root cause of our destructive presence on this living planet: population. It won't matter if we transition to all clean energy, all clean industry, local food, and net-zero buildings, if population keeps growing. It is not hard to see what a better chance we—and other life—would have if we could get down to a few billion less than we have now, and stay there. It seems that we often think about population in terms of "control"—draconian measures, such as China's one-child policy. This is a bar to seeing what is right before us—a history of lower population growth among developed countries.

This, then, inescapably leads to the thought that if, instead of a system of international trade that provided moneymaking opportunities for multinationals and wealthy investors, we had one that fostered domestic economic health and preserved environments and public welfare worldwide, populations would come down of themselves. When we help people to be secure in their lives, and make sure that women are free and empowered, we get lower population growth. It is by bringing prosperity and security to the desperate poor that we will reduce population. "The total fertility rate for OECD countries is below the replacement fertility rate."[18] *We have the information to know what to do about the problem of sustaining life on this planet.* But we face enormous psychosocial obstacles of ignorance, inertia, and lack of imagination and faith, as well as outright opposition to change. The fact that the United Nations (UN) has developed global goals to bring prosperity to all countries is helpful. If we continue the development of this thinking, knowing that it would be a way of turning down the primary source of our problems, that would be better than simply reacting negatively to the idea of population control.

Some attempts to grasp such a large topic are written by teams of experts, such as the demonstration of world culture that is the Intergovernmental Panel on Climate Change, or the rapid education in practical world saving that is Paul Hawken's *Drawdown*,[19] or the conjoined intelligence of the National Academies of Sciences, Engineering, and Medicine, which recently convened a committee to figure out how to decarbonize our energy system. Their first report says:

---

18. Salman Sakir, "How Low Population Growth Is Posing Challenges to Advanced Countries," *HuffPost Blog*, August 11, 2015, https://www.huffingtonpost.ca/salman-sakir/oecd-population-growth_b_7958124.html. The OECD is the Organization for Economic Cooperation and Development and consists of countries considered highly developed. As the low population growth will create the problem of a shrinking labor force and aging population needing care, Sakir recommends increasing immigration.

19. Cannot recommend this highly enough, as it concisely shows, with concrete specificity, what we can do.

> With an appropriate degree of humility regarding the influence of as yet unknown changes in consumer behavior, technologies, and other transformational changes, the committee has focused this interim report on no-regrets actions that will position the United States on a path toward a net-zero economy.[20]

No-regrets actions are things to do to slow climate change that are good things to do anyway. Or see the technical explanations of how we can get there by the Stanford University team headed up by Mark Z. Jacobson.[21] We *can* position ourselves on the right path. It won't just happen by itself. Many more need to see that we have the capacity and to exercise it.

Building on the information about our technical capabilities is the Environmental Law Institute's *Legal Pathways to Deep Decarbonization*,[22] by Columbia Law School's Michael Gerrard and Widener Law School's John Dernbach, with chapters by fifty-nine legal experts. It contains model and actual laws that can be downloaded and tailored to effect the change we need, to stop emitting greenhouse gases from transportation, buildings, power generation, and industry. The scientists and the lawyers are telling us *we can do this*, and they are laying out road maps.

EMOTIONAL CONTEXT AND POLITICAL WILL

Political will is an amalgamation of many individual wills. As a society, on this and many similar issues, we behave on the aggregate like cigarette smokers who can't quit. Ask any former smoker how easy it was to quit. Chances are that they knew they should, and may have even hated that they smoked, but it was still hard to change. Perhaps the bit of self-revulsion that smokers feel is in order. Lots of things ought to give us a sick feeling, as a polity, so that we act to reduce the level of insult to our bodies that poisons bring. Disgust may cause abreaction as well as a positive outcome. The emotional context of our decision-making requires attention, for we have choices we can make.

But it is not going to happen unless many more pay attention, many more *think* we can do it, and many more get involved in using all the tools in the great toolbox of governing options that we have. There's an old, stodgy, stuck

---

20. *Accelerating Decarbonization of the US Energy System*, 2021, https://www.nap.edu/download/25932 (download a pdf for free).

21. https://web.stanford.edu/group/efmh/jacobson/. For example, the road maps for 139 countries and all fifty US states to transition to all-renewable energy by 2050, 80 percent by 2030, written testimony in Congress, 2015: https://web.stanford.edu/group/efmh/jacobson/Articles/I/15-11-19-HouseEEC-MZJTestimony.pdf.

22. https://lpdd.org/pathways/.

story about government—the plodding and dull bureaucracy. But there's another government that is creative and innovative. It needs the support of environmental citizens. It needs enough to provide livings to the many young people eager to make a difference, who could pick up the good work that has been done and develop it further. Government can provide jobs in agencies that provide environmental assistance, it can provide grants to nonprofits, contracts to professionals, and it can set standards for the goods and services it purchases and post bonds for the development of the systems we need.

The job cannot simply be left to experts. They know their specialized areas, but this involves everything. Citizen involvement should be primary and expertise intended to assist us—the widest group consciousness possible—in identifying options and weighing actions and outcomes. It is citizens, united, participating not just out of self-interest but also to help protect the common—who should be primarily engaged in forming the necessary value judgments.

THE GOLDEN AGE OF ENVIRONMENTAL REGULATORY POLICY

There once was a time that I call the Golden Age of Environmental Regulatory Policy. For me, it began in the late 1980s, when a national consensus was building, to prevent hazardous waste generation and not just try to manage these awful wastes after they were generated.[23] Source reduction grew in the minds of planners and attentive policymakers until it was sanctified by Congress as pollution prevention in the 1990 P2 Act. The era included efforts to begin what I call relational environmental governance—an effort to understand what else we can do besides "command and control." There was an era of introspection concerning the relationship between government and the governed.

Neither of these movements got so far as to really pique the national mindset, and both unfortunately peaked at the end of the Clinton-Gore administration, under the Environmental Protection Agency (EPA) administrator Carol Browner, who oversaw four years of public discussion about how the system could be improved with a project called the Common Sense Initiative. The Common Sense Initiative, sometimes the easy target of ridicule by skeptics, was for me exhilarating and confirming. I saw people from all sides working together to find ways to achieve even better environmental protection without creating unnecessary adverse economic consequences. To me, it is a great

23. Massachusetts held four national conferences on "source reduction," and state after state reacted to the growing volume of hazardous wastes by launching programs to reduce it "at the source."

tragedy that this worthwhile project has not been understood as a powerful example of how to break through the binary division and find ways to progress, to build that more adult, more capable society.

The hard work of the reinvention era was then discarded[24] by the next administration, but not without some co-optation, so that some of that relational era became seen as being too friendly with polluting cronies,[25] confusing what happened to it and how it has been abused with its deeper purpose.

Hundreds of suggestions were developed and many successfully implemented, but the impact of bringing people from many perspectives together, and having such constructive results, has been lost. The impulse of that time, that produced real knowledge about how environmental regulation could be vastly improved, can be revived.

The environmental agencies may have been defunded, but they are being refunded, and the environmental movements have been growing all this time. People are fighting toxics in playing fields and lead in homes and algae growths and asthma-causing exposures, as they always have, and getting better at it, and they are bringing composting into schools and farms to urban lands and creating community land trusts and community solar. If you remember the *Whole Earth Catalog*,[26] those people who were taking the creation of a better world into their own hands never went away, and there are more of them, and more kinds. With good intent and fellowship, they are living a new way of relating to the mass of others we must first learn to live with on this planet if we are to learn how to live with life itself.

Laws are tools for doing that which we own—our self-organization. We can take them out and recalibrate them, reshape them, and make them work well. They have been captured and twisted to some extent, but there is a basic soundness in them and we can recapture and untwist them. But first, we must untwist ourselves so that the body politic can act to protect itself.

It is necessary to recognize the failures of the past and their tenacious persistence, how this seems to predict an inflexible outcome each time the vision of something better is raised. But in the end, it remains true that it is a kind of failure in thinking that, because of those failures of the past, there is no chance for a better future. It abandons the connection to our potential. In this moment of planetary peril, we need that connection. To demand proof that this hope will work before adopting it, is not how hope can work.

---

24. Quite literally as in documents being lost or recycled, websites removed, and initiatives discontinued.

25. One example was the Design for the Environment environmentally preferable product certification (now Safer Choice) established under Clinton, which lacked real verification under Bush (then put in place under Obama). Another was the Performance Track program to reward environmental responsible companies, canceled under Obama because of lax provisions, https://www.greenbiz.com/article/epas-cancelled-performance-track-program-finds-new-life-nonprofit.

26. https://archive.org/details/wholeearth.

The world society of problem-solving citizens is young. Think of it walking in many places, among the decaying monsters of the past: solar quietly replacing coal. Think of how the indigenous cultures of the world have gained recognition in international treaties, have had meetings amongst themselves, have begun to see laws passed to protect their lands, how marvelous an evolution this is from raw colonialism, and at the same time how murders of indigenous leaders accelerate.[27] We live in a time of great advance but we are still quite primitive in many ways and places, and keep showing how much more primitive we can get. There are forces that rise up to contest the advance. The world is crowded and the reconciliation that is upon us is resisted in a myriad of ways.

We can keep before our eyes the idea that the government can be tasked with bringing us together to solve our common problems. When people can see how well this can work, perhaps many will abandon the current trend of destructive, divisive politics for a constructive mode of interaction. Form a group and invite your legislator to talk. Get to know the planning board. Sample the water in the local creek and send in the results, along with lab certifications, when they are troubling, and communicate with the relevant officials. Join the conservation committee, help take care of the parks, bring food to shut-ins, bring old clothes to the homeless, grow vegetables and give some to neighbors, and, by these and other acts of interaction, create and know community. It is out of a life lived in relationship to others that a sensible system of governance is built, that which can enable us to transform our systems of production and living.

It's important to have a plan. But to paraphrase a celebrated book about the Constitution, it's not a machine that goes by itself. We need to be in it, making it work.

THOUGHT EXPERIMENT

Imagine that you are given the task of uniting the community around the goal of having clean water. What do you do? What do you say?

Any other environmental or public health issue can be substituted. Thinking through such an issue by yourself is difficult, but it trains the mind to consider how to change people's minds and behaviors. This is an aspect of environmental work common to most problems we face. How do we get consumers to buy safer products and companies to make them? It's not just

27. For example: https://www.culturalsurvival.org/news/cultural-survival-condemns-massacre-indigenous-leaders-colombia. See also the legal empowerment project of Namati: https://namati.org/what-we-do/.

chemists in a lab, it's advertising, it's quality of performance, it's testing, it's understanding the life-cycle of the product, but, above all, it's psychology and economics. We need the consumer to want the safe product and demand it—that's psychology—and we need the economics of safer product manufacturing to work. Technology and science—both industrial and biological—are involved, and the law can be used to change the economics, as well as the psychology. What tools of the law would you apply to get the change you need?

Thinking about this improves our ability to understand the position of government, and to know what to ask of it. If environmental citizens do thought experiments like this it will prepare them to help government discover how it can best serve the needs of the entirety.

Write out your own plan for whatever it is first, and then share it with the group. Compare this procedure to simply opening up the conversation, which may not result in everyone hearing all of everyone's thoughts, producing a diminished totality of concepts. Make sure everyone feels safe sharing their ideas—that people are not uncivil or critical. Critical analysis can come later, after we first get everyone's ideas out there. Make sure people understand that they are not expected to have answers but may write their ideas out completely in the form of questions. The assertions they make should be understood as general, potentially general, or specific, and an effort should be made to understand their factual basis. Discuss with everyone the quality of information, sources, and credibility.

Chapter Two

The Government of Us

Reinhold Niebuhr asked in the 1932 *Moral Man and Immoral Society* how it is that individuals can be so good but, as a group, could abandon their values and act awfully. By doing a better job of using democratic governance, we can bring about a greater reflection of that individual tendency for goodness in the group.

Each individual living on this planet now has two identities—that of their individual lives and concerns, and that of a member of the human species. It may seem like a big evolutionary, socio-psychological, cultural step to take, to integrate those two identities. But the environmental storm is forcing this recognition. We are inescapably being contextualized as members of the entire world and not just of our individual worlds. We are members of the human superorganism and each have a responsibility to the parent world that holds all of life. All future humans will look back at those of us living in this time and wish we had done more than we are now doing.

We always have before us two kinds of ethics: one in which we try to learn to live together and one in which we fight for our own, against others. The first is loyalty to the universe and the second is loyalty to family or group. The evolutionary step we need is how to observe both. It seems to me that societies need to put some work into cultivating the first ethic and altering the second, so that we each fight, as we must, to protect our own but against others only when there is no other way and not as a starting premise, as is all too often the ideological stance that is followed. And that we see we own, together, responsibility for the world.

The ethic of universal identity helps because the environmental emergency causes the pain of the recognition of how bad things are, and the idea that one is not alone with it brings some consolation and, if developed, real hope.

It is important to dwell for at least a moment on the pain that the loss of nature causes. Stifled or unrecognized environmental grief is not healthy. Many feel what Henry David Thoreau felt:

> When I consider that the nobler animals have been exterminated here—the cougar, the panther, lynx, wolverine, wolf, bear, moose, deer, the beaver, the turkey and so forth and so forth, I cannot but feel as if I lived in a tamed and, as it were, emasculated country. . . . Is it not a maimed and imperfect nature I am conversing with? As if I were to study a tribe of Indians that had lost all its warriors . . . I take infinite pains to know all the phenomena of the spring, for instance, thinking that I have here the entire poem, and then, to my chagrin, I hear that it is but an imperfect copy that I possess and have read, that my ancestors have torn out many of the first leaves and grandest passages, and mutilated it in many places. I should not like to think that some demigod had come before me and picked out some of the best of the stars. I wish to know an entire heaven and an entire earth.[1]

I too would like to know "an entire heaven and an entire earth." But there is an absence of mourning for the world, now so much more diminished than it was in Thoreau's time, already shorn enough of teeming life to prompt sympathy for a "maimed" nature. If you grow up without the song of birds, you may not know what you've missed, but that tragedy can be rectified—programs can preserve it and bring you to it.

There is an understandable fear of taking a good look at the emergency—and at us. It could mean not just our own personal death, but many more—if global warming releases the methane in the tundra and we get an unbearably hot planet, if a nuclear war kicks up dusts that block the sun for so long that everything dies, if insects and birds and trees keep dying, or if ecosystems simply crash—the extent of suffering in these scenarios amount to the mortality of us, and facing our own is hard enough. As psychoanalyst Stephanie Brody writes in *Entering Night Country*:

> The activation of mortal awareness, that our death is a non-negotiable contract with no escape clause, is unlike any human awareness. . . . The affect associated with anticipated extinction is, I think, beyond measure. . . . When Freud argued that "no one believes in his own death . . . in the unconscious, every one of us is convinced of his own immortality," he spoke to the context of timelessness within unconscious processes, the rejection of transience.[2]

---

1. *The Journal*, entry March 23, 1856.
2. Subtitled *Psychoanalytic Reflections on Loss and Resilience* (2016), 82. The quoted Freud is from "Thoughts for the Times on War and Death" (1915).

There are obstacles to overcome, but they are not primarily technical. They are primarily political, and they have their roots in our thinking. Consider the behavior of former President Donald Trump concerning the COVID-19 virus. He tried to wish it away, used hand gestures to suggest it would simply disappear, failed to take action. Hundreds of thousands may have been saved had we acted. Facing our problems is better.

ACTIVE CITIZENSHIP CAN SAVE US

The world is approaching "tipping points"—think of a cliff we fall off—places of extreme danger from which we cannot turn back.[3] Because the problem is so immense, so terrifying, we wonder: How can we, as puny individuals, tackle it? This question, faced, brings us to the same place: we are better off facing this together.

Civic leaders call for more engagement and periodically we do it, but not enough. We do have a national law that provides for civic participation in rule-making—the Administrative Procedure Act of 1946. It is important for citizens to know it and use its provisions. But it needs upgrading. The National Civic League's Working Group on Legal Frameworks for Public Participation proposed amendments to enhance involvement in policy-making, including a specialist position that "coordinates community engagement and trains agency personnel in participatory practices."[4] We can have these or other improvements by paying attention and wanting them.

However it is won, once begun, cooperation can give birth to a realistic optimism, which generates confidence in the context. The cooperative context is a very different interaction than the transactional. Think about the positioning and maneuvering one performs, witnesses, or must suspect in a purely transactional relationship. Or think about a context in which expertise is disconnected from purpose.

The cooperative context enables the combining of effort for shared purposes and the widest and most equitable sharing of resources and benefits. Government is the neutral, objective, universal party that can appropriately initiate cooperative effort. This can become a prime tool for us to ask all levels of government to step in, fill the gaps, and provide alternatives that are

3. For example, the melting ice of Greenland creating a cold "blob" in the Northern Atlantic that disturbs the Gulf Stream, which could radically change the climate of Europe and Northern Africa. Or the release of methane from a melting Arctic rapidly accelerating climate change. Or, most horrifying of all, the runaway greenhouse that keeps getting hotter till the Earth is like Venus. I hesitate to include this terrifying vision, but we might be at a point where there is motivational utility in entertaining such possibilities.

4. DeCaro et al., "Legal and Institutional Foundations of Adaptive Environmental Governance," *Ecology and Society* 22, no. 1 (2017): article 32, 13.https://doi.org/10.5751/ES-09036-220132.

responsible and equitable for consumers, workers, and producers, to be part of or partner with, instead of exploitative enterprise.

A society made up of autonomous entities responsible to care only for themselves does not produce the overall protective context the Earth's ills are revealing as necessary. Good mutual self-governance comes from a sense of community. Providing the basics of community and the interactions that form it must be a fundamental purpose of government in a self-governing society. We need not let this function of government wither on the vine from disuse but, instead, can turn to it now as a centralizing function.

The concerned citizen has a lot of places to turn to find out how we can address the great, looming disaster of climate change. (For a ten-point plan for a low-carbon future, see the World Resources Institute).[5] But the plans of the experts will not get us far enough unless the rest of us pay attention and think about what they are telling us, and then add our voices to a world-spanning conversation that transcends divisions and generates the sense of shared presence we need.

We need worldwide government of, by and for the people of the world.

To solve our problems, a change in the prevailing attitudes toward government is necessary. One can be horrified by some things one's government does—such as dropping napalm on villagers in Southeast Asia and such as police brutality—without becoming cynical about the idea of democratic government. A country, such as the United States, is both a beacon of hope for a just, democratic society and a hegemon with hundreds of military bases around the world, supporting an unfair, exploitative global economic system. We have to sort things out: untangle from the latter and build up the former.

In *The Dwarf*, by Par Lagerkvist, the narrator observes Leonardo da Vinci and marvels that he can hold two very different, seemingly opposing, thoughts in his mind at once. With this concept, we can stop trying to shove our thinking into either one box or another, and we can all inhabit the large box that holds all of us. We can hold dualities in our minds and experience a multidimensional sense of reality. We can consider points made by Democrats and Republicans in terms of their own merit, not by which team expressed them.

Expansion of identity—acknowledging the presence of others' realities and responsibilities beyond one's small world—does not threaten who we are. It simply opens us to the ability to relate. We belong to so much together that the end result of authentic encounters between people of goodwill is likely positive. Removing the barriers to this natural state, removing the encrustations, the unnecessary encapsulations, the limits, that have put us into

5. https://lpdd.org/wp-content/uploads/2020/04/WRI15_WorkingPaper_post-2020_0.pdf.

warring camps, no longer deciding before trying that something we desire is impossible to have, we can regain the ability to turn things around.

Nothing about our failures tarnishes the magnificence of ideas embodied in our systems—now in the constitutions of governments around the world.[6] Thomas Jefferson enslaved human beings, gave power to the idea of equality, and worked to separate church and state and promote science. Each of us is a mixture: nothing about our failures requires us to abandon hope: just to work more effectively to develop.

Though the organs of democracy—the legislatures, the agencies, the courts—have been corrupted by narrow selfishness to an alarming degree, they are still the most magnificent creation of human civilization, and belong rightly to us. They still have many people in them of integrity, and faith that they can be recaptured for use, for the people, lends energy to the task. Asserting that we can take them back increases the chances that we will take them back. Then we can implement the excellent suggestions of the National Academies of Sciences, and all the other experts who have told us how to save what's left.

GETTING THE STORY STRAIGHT

Some who harm the planet don't do it intentionally, and may see no other way to get along. But many do have other choices, and instead of choosing responsibility, they have chosen to work to make our government represent their selfish impulses. It is reasonable to plan to use government to create disincentives for this kind of behavior. There is a difference between these two groups. One rejects the idea of sovereignty of the people, essentially usurping the central position, and the other is just trying to get along and would likely willingly accept a reasonable deal that provides a way to get along without unnecessary harm to others. There is, in addition, a growing business sector that is managing to produce things people want, without the harms. It is up to society to distinguish between these groups, and it has the opportunity to structure incentives and disincentives appropriately, with the goal of gently and carefully evolving more responsible life practices.

In the *New York Times*,[7] Nicholas Kristof talks about how so many people he knew growing up—loggers—have suffered because of the loss

---

6. See the 2015 *Global Environmental Constitutionalism*, by James May and Erin Daly. The authors note a rapid growth, from about fifty constitutions that explicitly recognized a fundamental right to a quality environment in the mid-1990s to about ninety-five constitutions with such provisions in 2012 (4).

7. "Can Biden Save Americans Like My Old Pal Mike?" https://www.nytimes.com/2021/02/13/opinion/sunday/working-class-dignity.html

of jobs, and how blaming the environmentalists, who stopped the uncontrolled cutting of forests to save owls, became widespread. Somehow the story has gone out and held fast that environmentalism is bad for people. Maybe if our society had worked harder to help the loggers, the resentment of environmentalism might have been avoided. But also, political opportunists and propagandists working for polluting industries have seen an interest in promoting the idea that liberal environmentalists care more about animals than people. This is an unfortunate part of modern environmental policy history.

We may not shed a tear for the wealthy owners of coal companies who lost money when coal plants were shut down. But if the story is about help for the miners, who have also needed help while employed (stricter safety inspections, for example, would have prevented much injury), that might change things, as well as more widespread recognition that their loss of jobs was mostly due to mechanization and anti-labor actions, not environmentalism.

If the story that environmentalists are to blame for lost jobs is what people hear, we cannot expect support for environmental initiatives. Environmentalists may prefer nature and animals to humans, but they should remember that we must develop widespread political will for effective environmental governance. Environmental law has to be seen as good for people. Helping people as a visible and indeed primary priority will be necessary if we are to succeed in winning people's support.

I have found myself thinking in a "biocentric" or "ecocentric" way, and I consider it a blessing to have a glimpse of the world from a perspective that isn't so limited by purely "anthropocentric" viewpoints. It is an illusion, though, I feel, to think that in the end I can ever see things in a way that a nonhuman would, for no matter how I think about what I see I am a human seeing and thinking. But the sense that all life-forms have value, that life is all around you, is a connective energy to be sought and treasured. It is not in conflict with caring about people first and foremost, but enhances and invigorates it. An effective anthropocentric approach *necessarily encompasses protecting the context in which people live and thrive*, which is one that protects all life-forms.

When stories pose conflict, momentum stops, but stories can be reframed. Stories that reveal the congruence in bio- and anthropogenic attitudes are uniquely helpful in renewing momentum.

A STORY ILLUSTRATING HOW IT IS NOT FOOLISH TO HAVE HOPE

In the early 1990s, when I was working at one of the environmental agencies of the state of Massachusetts, in a new program that was set up to provide assistance to companies, rather than enforcing against them, I received a request for help from a POTW. A POTW is a Publicly Owned Treatment Works—a sewage treatment facility. POTWs handle the wastes that people flush down the toilet. They also receive wastes from industry, which can be a problem for them. Sometimes what a company sends down the pipe is not something the POTW can handle. This POTW was violating its own discharge permit because of what companies were discharging to it. In this case, it was copper. The POTW was in an area with a lot of textile companies, and they were discharging a lot of copper, which was in the blue dyes that they used. It was too much for the POTW to manage, so it passed through into receiving waters, and EPA had told the POTW that it was violating the Clean Water Act limits on discharges of copper. So, the POTW told the companies that they had to stop discharging so much copper to it.[8] But the companies were telling the POTW that there was nothing they could do. The POTW kept fining them but nothing was changing except the companies had to pay money, and they were facing stiff economic competition from other parts of the world. The POTW reached out wondering whether my office could help, so the companies wouldn't all leave town and relocate somewhere with lower operating costs.

I read up on what I could find about textile dyeing and I invited the companies to a meeting to discuss the matter with me and one of my colleagues. Because we represented the state, and because the POTW had enforcement power, they came. They told us just what they told the POTW: there was nothing they could do about it. We talked for a few hours and got nowhere. The copper was in the dyes they used, and they had no control over that. They had to operate as they always did, or they would produce poor-quality goods, and this was a very tightly competitive business. I remember them talking about how they were earning pennies. They worked hard to make us want to leave them alone, as if the problem was just going to go away. They leaned heavily on the issue that they might have to relocate.

Back at the office, I got on the phone with the companies that make the dyes, and they said the same things, but agreed to come to a second meeting. We presented the problem again, told the dye makers that the companies, their customers, had told us they were unable to deal with the problem, and asked if there was any way they, the dye makers, could help. At first, I thought it was going to be another wasted effort, for the dye makers said they

8. In policy jargon, the POTW is enforcing "local limits" on an "indirect discharger."

had to make the dyes the way they did. Copper was essential. You couldn't get rid of it. Alternatives didn't do the job.

But at that second meeting, a kind of miracle happened. It was one of those things that makes you want to keep persisting even in the face of seemingly insoluble problems. One of the textile company chiefs suddenly said, "well, you know since our meeting I've been thinking." He then explained that if you slowed down the rate at which the textiles were run through the dye bath, you could increase the uptake of the copper in the bath on the cloth, and that that actually reduced what was left in the rinse-waters. I well remember how the other companies looked at him: it was like he was a traitor. *Slowing down production was going to cost money and we are operating on penny profits!* But then, and this is a continuation of the miracle, something seemed to melt away, and we heard more ideas. My colleague was an engineer, paying attention to the chemical details. I was paying attention to the personal dynamics. Then my heart leaped—one of the *dye makers* started talking about changes he might be able to make in the dye formulation that might help the copper bind more tightly to the cloth. I felt like we had blown a ram's horn and the walls of Jericho had come tumbling down.

A short time later, my colleague[9] won us a visit to a textile dye factory. Remember that this whole thing had started with adamant insistence that "there is nothing we can do." When we walked into the area where the copper dye was put into the dye bath, I saw a worker plunge a scoop into a big bag of dye granules. A number of them spilled over onto the ground as he transferred the material. I pointed at that and asked what would happen to it, thinking that the answer, of course, would be that it would be swept up, maybe recovered and used. But no, the answer was, "we hose down the floors." That hosing was an avoidable discharge of copper down the drain. With many scoops and spills per day, it was significant. This was pretty far from "there is nothing we can do"! It didn't take long for changes like that to get everybody in compliance with the Clean Water Act, and we now had a working relationship with a community of factories who trusted us to help. Government was no longer the enemy of business.

My job at the state, for two and a half decades, was to help companies reduce their toxic impacts. The claim that "there is nothing we can do" was not something that only textile companies said. Every industry category said that. Almost every dialogue I initiated with industry began that way. But the number of times that I saw simple, obvious reductions like the hosing down cannot be counted. They were everywhere. There was a great deal that could be done.

---

9. Rich Bizzozero, who later became director of the office.

The fact is, industry doesn't like to be told what to do, and there is a natural inertia that keeps you doing what you've always been doing—this is especially strong when you know it will get that product to market and let you make payroll and keep the wolves from the door. This is all totally understandable. The system of environmental protection that we have put in place, which has saved much of the world from premature destruction (still happening, but it could have been far, far worse), relies primarily on telling industry what to do. It's referred to as "Command and Control." It has to be done. We must not tolerate the poisoning of our world. However, note that we used a different approach. We didn't give any orders. I just asked questions. Have you looked at this option? Call it "Options Analysis." Note that it worked in this case: it doesn't always. You have to create the conditions for it work.

Had we not intervened, EPA would have fined the POTW and the POTW would have turned around and leaned on the companies even more. Maybe money would have been spent on lawsuits. Maybe they would have shut down and left town.

This story illustrates what has happened with environmental protection law. We put a system of environmental governance in place and have been fighting immense and costly battles ever since. This is one way to go about things and it had to be done. But there are two ways, and we need to do both. There are many people in industry who need enforcement. Some lack the sense of responsibility to others that restrains the unmitigated profit motive. And although many do have a sense of responsibility, the corporate structure, the need to please investors, and the pressure of daily production too often works against their moral instincts. While there are notable exceptions, the competitive economy is a context that sets up environmental protection for failure. But then there are opportunities to change even though there is so much that resists change, and we have to actively look for them. We have to create them. We can't just rely on the adversarial approach.

AN EXAMPLE OF HOW OUR PROBLEMS ARE ALSO OPPORTUNITIES

Our problems are also opportunities. On December 28 of 2021, a year in which a great mass of consumers of bad information generated deep anger toward our own system of government, slowing the repair to systems that had been damaged by the previous administration, the *New York Times* reported on the tension building between the need for rare earth minerals and the need to restore the rivers and lands of the West. The system of international trade, recently of attention because of COVID-caused supply chain disruptions, but for decades the cause for growing concern as it impacted local economies and boosted pollution, resulted in China's domination of the raw materials we need for our clean energy future. Here is a problem of great complexity, also weighed down by the baggage of bad policies of the past. But there is yet great opportunity: might we begin to rebuild local economies that have been adversely impacted by the growth of international trade?

We can use a framework for looking at all our problems that is neither predictively optimistic (just have the right outlook and things will be fine) nor cynical (deflating the political energy to move). The *Times* tells us, referring to a proposal to mine antimony in the Salmon River Mountains, that it has been established that the proposal would cause "disproportionately high and adverse impacts" on an environment once full of life, such as the salmon the Nez Perce tribe is trying to bring back. Here is an issue of great complexity, and seemingly no obvious light:

> The Nez Perce are not alone. Across the American West, tribal nations are on the front lines of a new debate over how to balance the needs and costs of clean energy . . . just as fights over the environmental costs of oil exploration helped define the fossil fuel era, conflicts like this one are creating the battle lines of the next energy revolution.[10]

The idea of this book is to sit with such issues and consider them, without jumping to conclusions, without consulting the stance of a party. We can begin with questions: Do we need to be having such conflicts? How does the framing go from debate to battle? What are processes by which we come to our opinions, take sides, and make decisions?

Readers of the *Times* story may only see that society has one choice—either inflict more harm on the salmon and the Nez Perce or give up trying to make batteries here. It reminded me of a debate that recently happened in my town.

The sustainability committee on which I participate was approached by a company that wanted us to sign on to participate in a large solar project. Members of the committee were very excited that we could lock in solar power for decades and rapidly accelerate our progress toward our net-zero goals. I asked if any trees would be cut down, and the answer was yes, ten acres. Worse, they weren't in our town. Eventually my opposition to the project won over others, who felt that it was not right to take another town's trees. I am a committed solar advocate, but I opposed this project because I know how valuable trees are, not just for carbon "sequestration" (the best device we have) but for so many other ecological benefits.

---

10. Jack Healy and Mike Baker, "Clean-Energy Gold Rush Alarms Tribes in US," *New York Times*, December 28, 2021, 1.

The solar proponents argued that trees would provide much less of a carbon benefit. They provided figures that showed we would get more carbon reduction from the solar panels. Studies showing this were shared by committee members. One of the town committee members presented a proposal to cut down our own trees, in an area behind the town composting site, surrounded by swamp, which I suppose he thought no one cared about. But I sometimes go there when the ground is frozen and said I would miss the trees there if they were cut down.

I told this story to my students and asked if anyone was interested in thinking this through with me. How could we do solar without having to cut down trees? Because if a house was burning, and we had two hoses, would we say, hey, let's just use the bigger one? Students have taken up the issue (and presented to the town committee). You can put solar panels by the highway, over parking lots, bike paths, and much remaining roof space. There are fun designs you can use to place panels in parks and public spaces, and increase plug-in opportunities.[11] As the summers get hotter, we will all want more shade, so why not construct walkways that double as solar panels? (Make towns look like the Old West). We could surely use them in public transportation to cover outdoor boarding. What about all that unused space under powerlines? And since some solar panels allow light to pass through, some are experimenting with growing things under them. We can place panels vertically on south-facing walls. An era of "novel solar placement" beckons. There may be reasons why some of these ideas won't work, but if any do, we can save trees. Instead of pitting good strategies against each other, we should find ways to use all of them at the same time.

Sitting with complexity is no promise that ideas will come. But think about how often we see that, as opposed to the instant drawing of battle lines. Just feeling the tragedy of the salmon's decline, and the awfulness of not having the ability to build batteries can produce despair, anxiety, and depression. But if, instead of turning away, we actually keep thinking, we may eventually realize an important fact that should change how we think about whether we mine in areas that destroy ecosystems: *we are not recycling the batteries we have.*[12] We are not recovering these materials from the waste streams of the products we put them into.

---

11. See www.bu.edu/rccp, in Summer and Fall 2021 and Spring 2022 projects.

12. Perry Gottesfeld, executive director of Occupational Knowledge International and a member of the California Environmental Protection Agency's Lithium-ion Car Battery Recycling Advisory Group, recently wrote that "lithium-ion battery makers have yet to develop the technology that can economically extract components in a form that can be used to make new lithium-ion batteries. Rather, the batteries are typically processed to remove the cobalt and a few other expensive metals, with much of the remainder released as air emissions or used as filler in concrete or other construction products. This is one reason why less than 5 percent of lithium-ion batteries are currently recycled," https://undark.org/2021/01/21/electric-car-looming-recyclability-problem/. He notes that only three states have acted to require extended producer responsibility concerning vehicle batteries.

We may ask: Why must raw antimony be mined? Can we not recover antimony from the many products in which it has already been incorporated? I do not know the answer, but the *Royal Society of Chemistry's Periodic Table* page tell us that:

> Antimony is used in the electronics industry to make some semiconductor devices, such as infrared detectors and diodes.
>
> It is alloyed with lead or other metals to improve their hardness and strength. A lead-antimony alloy is used in batteries. Other uses of antimony alloys include type metal (in printing presses), bullets, and cable sheathing.
>
> Antimony compounds are used to make flame-retardant materials, paints, enamels, glass and pottery.

This makes me think: How about we incentivize "mining" companies to go looking at the waste streams coming out of all these uses? And let the Nez Perce bring back the salmon. This is neither to say that the answer is simple nor that this idea settles the issue. It is to say that we need to spend time trying to be creative about problem-solving and do that famous thing of thinking outside of the box.[13] It is something we can choose to have as a common goal and activity. If the "front" of our struggles shifts from destructive, time-wasting fights between people to the problem itself—a shared thing—we could employ a cooperative effort to produce a more holistic view where we see the problem more clearly and have the chance to find the solutions of highest optimality (combining justice and efficiency, as elaborated below).

I don't believe that we need to be thinking about all of our environmental problems as conflicts. Many are, and there is needed strong resistance to damaging activity. But I do believe that if we, as a society, accept the goals that Mother Earth is screaming for us to accept, then that means we must more quickly develop intelligent rules that shift economic incentives toward doing the right thing, and we cannot develop those more intelligent rules unless we deal with each other intelligently.

The *New York Times* article reminds us of the history of fighting over oil exploration. That and other fights over environmental matters all stem from the mistaken idea that we can afford to damage the Earth, and we can't afford to get in the way of what financial investors want to do. It's the other way around. If we as a people unite around the idea that we do have responsibilities toward the environment, and take them seriously, we can

---

13. To think outside of the little box of a small view of the world, that is. We need to think *inside* the box of the biosphere's requirements for stability and health.

find ways for businesses to succeed. We must defend the world, including the truth about the world. That means when the things we love are attacked, we are in battle mode, whether we like it or not. But if that is all we do, we will lose the chance to forge a new consensus. We *also* have to make efforts to create the shared space that can transcend conflict. Each of our environmental problems calls for creativity. We don't know yet how we will solve them, but we will sooner have ideas if we cooperate in seeking creative solutions.

IT IS NOT TRUE THAT WE CAN'T CHANGE

We can change how we go about environmental protection. The system has suffered from a prolonged onslaught and in some areas has become quite ineffective. Much public support has been lost because of the propaganda campaign against environmental agencies, and we should act to counter those immoral campaigns. But there are reasons those efforts work: the system needs not just to be updated but to evolve.

Let's imagine that, in addition to enforcing environmental standards, we also put a lot of effort into trying to help industry to make the changes we need them to make (less pollution, less waste, less risk, less damage, etc.). If the assistance doesn't do the trick, then we use enforcement. But if assistance works, we will reduce the problem, make friends instead of enemies, and save on the cost of enforcement. It makes sense to try assistance first. Lead with the helping hand and keep the whacking one in reserve, as a general rule, tailored to each situation, as necessary.

My career gave me experience with both arms of government. I had the job of calling companies and saying we were seeking fines for violations, and I had the job of calling companies and offering the hand of assistance. The program of assistance where I worked was astonishingly successful, and it is by no means a rare example. But it was for a particular population. The program of enforcement was very efficient and appropriate for most of the population I saw addressed by it. But many of those caught up in it could have been kept out with the right assistance program.

Environmental agencies have long engaged in education, assistance, guidance, and other friendly activities. But these efforts are not the primary face that people see. Far too many fear and resent environmental agencies. Far too many try to hide from them. Far too few pick up the phone and call the agencies and ask for help. A real effort is needed—a visible and highly supported investment in bringing a complementary assistance effort to the fore—to counter negative perceptions of the relationship between environmental agencies and the public.[14]

Sometimes we can set absolute standards, but determining the final answer to what is safe might take too long, doesn't change easily with the times, and is something to fight about. If we think about relative risks, we can focus on the worst first, and use the principle of reducing risk where we can, to the extent that we can.[15] This is a way forward that can be applied to any context. We need strategies that can be quickly understood and can't really be denied,[16] and maybe they can survive changes in administration.[17] With general consensus, we can have a government of us.

CLIMBING OUT OF OUR HOLE BY GRASPING THE WHOLE

Because we tend to get focused on particular things at a time (because of prioritization, constraints, distractions, mistakes, visibility, political pressure, mandates, etc.), it is necessary to look at the big picture. We focus on air, water, waste, worker exposure, toxins in products, and spills and cleanups, one at a time, sometimes without seeing how they relate (many times all pointing back to the same decision to use a cheap toxic material). We focus on one place at a time—a particular site, a state, or a country. We need to start with the entire world and work down from there. If we don't base our programs on that—on the health of the biological context in which we live—they cannot amount to sufficient protection, nor does government

14. I believe that most of the public appreciates the important work of the environmental agencies, but *enough* of the public has accepted negative views so that opponents could win reductions in agency budgets.

15. In "The Regulation of Toxic Pollutants under the Clean Water Act," *Environmental Law Reporter* 21: 10528 (1991), Tulane's Oliver Houck writes:

> Congress should fix several timetables, based on relative risk. Science may not be able to set absolute risk numbers, but it can identify categories of greater and lesser risk. Industry can obtain adequate lead time for the necessary process changes. Economists can identify incentives that will encourage industry to arrive on or ahead of schedule. And the engineers can, at last, design for the inevitable. To the extent that certain industries are participating willingly and even aggressively in voluntary pollution prevention and toxic reduction programs, a mandate this firm should only strengthen their hand, and lessen their competitive disadvantage. The critical elements of this alternative are that the mandate, lists, and schedules be made congressional, and therefore beyond litigation, and that they be made firm. When it finally becomes time to make capacitors without PCBs, paint without lead, and refrigerants without CFCs, industry finds the way. At which point, ambient levels of these toxins drop precipitously.

16. Such as "you should take precautions that make sense," which is hard to argue against.

17. The ethic of pollution prevention has survived changes in administration, but barely. We could use a revival of its common sense because it could inspire a rare universal agreement (for example, the 1989 Massachusetts Toxics Use Reduction Act, considered the strongest of the state pollution prevention laws, was passed unanimously). An ethic—call it universal risk reduction—that one is supposed to try to see when one's own actions increase risks, so that they can be avoided, just makes sense.

do the job that makes it legitimate—working for all of us. However, because we can never know enough about how everything fits together, we need to think big and with humility at the same time. Just protecting certain aspects of the environment, such as selected "ecosystem services," will help but is not enough. We must try, together, to approach the objective perspective, without assuming we have reached it. We can think of *the biggest interdisciplinary inter-subjectivity* we can manage—putting together all of our knowledge and perspectives—as the way to do this. The planet is calling out for help and we can gather in witness of it and each other.

In 2009, Johan Rockström, director of the Stockholm Resilience Centre at the time, organized an effort by prominent scientists to assess the "planetary boundaries" that humanity must not cross, if large disruptions and irreversible changes are not to occur that threaten civilization's ability to thrive. The group identified nine boundaries:

1. Stratospheric ozone depletion
2. Loss of biosphere integrity (biodiversity loss and extinctions)
3. Chemical pollution and the release of novel entities
4. Climate change
5. Ocean acidification
6. Freshwater consumption and the global hydrological cycle
7. Land system change
8. Nitrogen and phosphorus flows to the biosphere and oceans
9. Atmospheric aerosol loading

The latest version of how we are doing was published in 2015. This graphic (figure 2.1) allows citizens of the world to see how we are in serious trouble in several categories.[18]

An appraisal of this effort and the many comments it prompted was published in 2020 in the *Annual Review of Environment and Resources*. The authors opine that "More research is needed to explore implications of planetary boundaries for earth system governance and dominant paradigms such as economic growth" and that "Key research questions relate to the institutionalization, coordination, operationalization, and democratization of planetary boundaries."[19] In other words, not only do we need more science but we need to change how we govern economic activity.

18. J. Lokrantz/Azote, based on Steffen et al. 2015. For background and related materials, see https://www.stockholmresilience.org/research/planetary-boundaries.html. E/MSY = Extinctions/Million Species Year. BIL = Biosphere Integrity Loss.

19. Frank Biermann and Rakhyun E. Kim, https://www.annualreviews.org/doi/full/10.1146/annurev-environ-012320-080337.

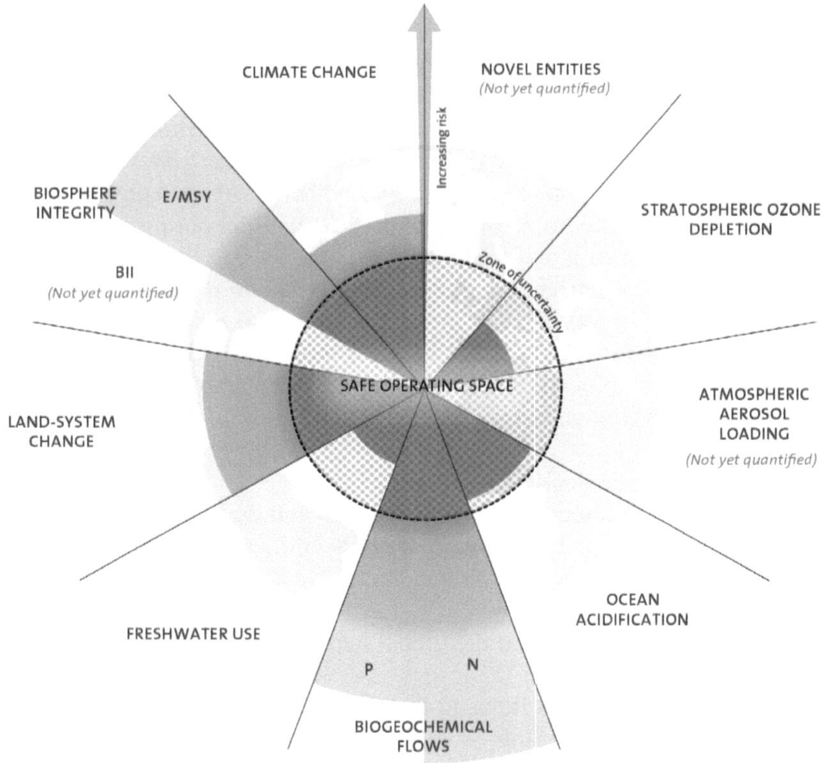

Figure 2.1. Stockholm Resilience Centre's Planetary Boundaries ***Source:*** J. Lokrantz/Azote based on Steffen et al. 2015. **Retrieved from** https://www.stockholmresilience.org/research/planetary-boundaries.html.

The issue of governance needs lots of attention now. A review of eighty articles about the planetary boundaries work noted that "the Earth system science-derived planetary boundaries framework describes the problem, but it offers little as far as solutions are concerned." In responding to the challenge facing the inhabitants of the Earth, "Earth system law must be fully representative and inclusive of all State and non-State interests (and their justice-related concerns), including of those in the global North and the global South, of present and future generations, and of humans and more-than-humans." The "Earth system law" we need "could meaningfully address the perceived democracy deficit in global environmental governance."[20]

20. Rakhyun E. Kim and Louis J. Kotzé, "Planetary Boundaries at the Intersection of Earth System Law, Science and Governance: A State-of-the-Art Review," *Review of European, Comparative, and International Environmental Law*, November 2020 https://doi.org/10.1111/reel.12383.

The scientists have explained the problem, which includes not yet having the political will to establish a system that transcends the state-mediated balance of power that international law currently comprises. They point to the task of establishing a world civilization that works. Much has preceded us which is useful in this endeavor. And much of what constitutes it is being created anew today. What remains is the will to affirm it, to use what we have created, to live the principles already at work. A worldwide effort involving and respecting everyone is the kind of civilization that has a chance of being ecologically healthy and socially stable. To assert this and make it real, would be humanity answering the call of its mother, the natural world, acknowledging its fidelity to what has nurtured us and what we love, and putting aside its lures, its instruments of torture, and its bombs and poisons, for a cared-for world.

THE TASK IS OVERWHELMING BUT LESS ONCE WE ENGAGE IN IT

Anyone wondering what they can do to help save this endangered world today should resolve to become more informed about governance, but this is not so easy for busy people who have to pay attention to many things during the day. Learning about what people in government are doing is hard enough, but to also learn enough to understand larger principles, such as how governments should be structured and interact? This seems impossibly complex. But moral principles can help cut through the complexity. Is government addressing our common problems actively enough? Are our leaders being fair to everyone? Are they ramming things through without discussion? Are they making crazy accusations of the other side? Many questions with complexity have simple, graspable points.

One seemingly impossible complexity is the socio-psychological questions of why whole populations act the way they do. The chart that shows we are already exceeding boundaries—grossly, too, concerning the flows of nitrogen and phosphorus, and extinctions—this should make us ask why these hugely important issues get so little attention compared to the latest news, sports, or entertainment. Why have we let ourselves get so far behind where we need to be? Answering that question, of course, involves history, culture, psychology, politics, law, the economic incentives for remaining narrowly focused, and many other aspects of who and what we are, so it seems impossible to grasp. But it is terribly interesting—indeed fascinating—once you dive into it, and complexity is not an excuse for being incurious. There is something to be said for the practice of regarding problems as puzzles, especially when you can win real benefits.

Consulting good sources of information is essential, for it introduces you to people who are solving the puzzles. It helps you see how things fit together—that is, making some time every day to learn about the issues that concern you ought to actually be able to lessen your anxiety. It may increase your concerns or reduce them. But you will, in any case, lose the confusion of ignorance. One is stronger with better information.[21] Foremost in importance is information that we actually can do what needs to be done, because it provides needed emotional and political energy.

In the February 2021 issue of *Scientific American*, associate editor Andrea Thompson wrote that the new President Biden could make substantial headway by focusing on transportation, power plants, methane emissions, and hydrofluorocarbons (HFCs). Regulations for fuel efficient vehicles, limits on emissions from power plants, stopping leaks of methane from drilling, cutting HFC use, tax incentives to expand renewable energy, a stimulus package to promote green infrastructure, government purchasing of clean energy, and making it harder to get approvals for polluting activities, could help the world catch up after all the time lost because of President Trump's opposition to responsibility. (Biden is pursuing all of these goals.)[22]

There is a big gap between science and action, which is usually filled by technology, which is often developed and employed for reasons of military defense or making money. But now we live in a new world, where anyone who cares about the future must see their hopes threatened. In this new world, clean technology is a shared, planetary need. Will we, or investors, determine what technology is developed? It is the right of the people to use their government to incentivize invention in the public interest and develop ways that the public can benefit from the investment of its assets.

The planetary boundaries idea has been taken up by an Earth Commission,[23] a project of the Global Commons Alliance,[24] composed of fifty organizations, and we also have an effort to save the world managed by the United Nations, the Sustainable Development Goals (which is more focused on helping people, but saving the environment is part of that), adopted by all member states in 2015.[25] They are orienting the world toward shared action. Even the

21. It is a tragedy that fine sources of information have been rendered suspect to many by being portrayed as partisan. Readers are urged to consider not just the sources they like but also why. For example, I believe the *New York Times* is far from perfect but that it makes a very great effort to tell important stories accurately. It corrects its own work constantly. It quite often presents issues in a way that reveals their complexity or significance in the light of other matters, ideas, or history.

22. A remarkable indication of the ability of industry to respond to the challenge was the announcement in late January by General Motors that it would cease making gas-powered cars, a dramatic contrast to the direction set under Trump, who rolled back fuel efficiency limits.

23. https://earthcommission.org/.

24. https://globalcommonsalliance.org/.

25. https://sdgs.un.org/goals.

person some regard as the most important investor in the world, Larry Fink of BlackRock, which manages $9 trillion in assets, urges businesses to plan for a "net-zero" economy.[26] When money moves to fund clean technology, there is hope. But that hope dies if the rest of us turn from the task, thinking the rich and powerful will take care of things for us. They have already shown a tendency to sequester too much to themselves. We cannot leave it to them.

SAMPLE PLAN FOR WORLD SALVATION

What would a real plan for world salvation look like? As an example, take the thought expressed above to begin with population and help families. In a 2016 article in *Nature*, Jon Bongaarts of the Population Council points out that "low levels of female education, a lack of knowledge about and access to contraception, insufficient supplies and services . . . (and) opposition from spouses and other family members and traditional gender roles that support a desire for large families" cause millions of unplanned pregnancies, but programs in Bangladesh, Iran, and Rwanda that provided assistance and education about the benefits of smaller families have been very effective. Major investments to scale up these approaches could lead to a billion fewer people in the already stressed sub-Saharan region by 2100 than is currently projected. He notes that

> Family-planning programmes are most effective where socio-economic conditions are improving. . . . Family planning, education (of women and men) and socio-economic development are mutually reinforcing, and should be pursued together.[27]

In other words, helping to develop local economies, and providing education and services, especially to help women take charge of their own lives, will help reduce the killing impact on planet Earth of too many people. The group Population Matters (which counts David Attenborough and Jane Goodall as patrons) points out that "The highest rates of population growth take place in the poorest countries":

> The world's poorest countries tend to have the largest family sizes and fertility rates. When people have no economic security and cannot rely on their government and a social safety net, they often have children to help provide income

---

26. Andrew Ross Sorkin, "BlackRock Chief Redoubles Sustainable-Stock Focus," *New York Times*, Business Section, January 27, 2021.
27. "Slow Down Population Growth," *Nature*, February 24, 2016, https://www.nature.com/news/development-slow-down-population-growth-1.19415.

or labour for their families, and to ensure they will be looked after when they are older.[28]

Valuing women's rights and bringing about global economic justice are thus critical first steps in saving the environment. Despite this fact, we have seen far too little investment in this effort, and it is hobbled by primitive urges to control women's bodies.

People can rise up and express their will to recapture government and make it work for them, and they must, before we lose the world we share. We need not and should not think of this uprising in terms of anger or violence. We can think of the Velvet and Singing Revolutions.[29] We can think of the Black Lives Matter marches, which, despite conservative focus on a few incidents, were in general very peaceful, as were the Pink-Hatted Women's marches, as was the majority of the civil rights movement. Our heroes can be those nonviolent organizers who showed how to reach the conscience of a nation.[30] Violence is counter-productive to the greater purpose and provides justification (though false) for repression.

The tools of governance are many and diverse, and they should not remain in an unopened toolbox. We should use them all, figure out how they can be used together synergistically, and develop new ones as needed.[31] The purposes of government, of and for us, can be manifested in programs benefiting our environment and health.

How we translate what science tells us about our problems into action that solves them is something we can all focus on. Without support, neither the Earth Commission nor the UN nor a 9 trillion-dollar fund will do it for us. But if we pay attention and work at changing out false dreams for real ones, we can write a new future. We need to know, when companies like Shell Oil are beginning to invest in electrical vehicle charging stations, not to say, *okay, that's great*, and go back to our own affairs, because they are only investing very little. We need to encourage them to continue changing and get going on incentivizing those who will really do the job as it should be done. That is good business and the public interest, and should be considered good governance, rather than the "slippery slope" to communism or a loss of freedom. In an active government that asserts the public interest in changing how we

28. https://populationmatters.org/global-justice.

29. Peaceful transitions from communist dictatorship in Czechoslovakia (1989, also known as the Gentle Revolution), and Estonia (culminating in 1991).

30. See, for example, Diane Nash's story, noting James Lawson, who learned from Gandhi, https://news.yale.edu/2017/01/27/diane-nash-urges-todays-activists-apply-techniques-civil-rights-movement

31. See, for example, Reibstein, "Using the Tools of Pollution Prevention to Reduce Greenhouse Gas Emissions," *Environmental Law Reporter* 38: 10851 (September 2009), https://www.bu.edu/earth/files/2012/08/Tools-of-P2-for-GHG.pdf.

produce modern life, is our chance to be free of many things we don't need to be suffering.[32]

EXERCISE

Identify the wastes you generate from all your activities. Consider what you know about the fate of that waste. Gauge the state of your understanding about how to reduce it.

32. "Shell, BP and Total toot their green horns while making relatively puny investments in clean energy—3% of BP's capital expenditures in 2019, 4% of Shell's and 8% of Total's" (Wal van Leirop, "In Denial, Oil Executives Start Peddling Filtered Cigarettes, Forbes, January 26, 2021, https://www.forbes.com/sites/walvanlierop/2021/01/26/in-denial-oil-executives-start-peddling-filtered-cigarettes/?sh=19ce522f1243).

Chapter Three

A Story from the Front

In 1990, the president of a small company rose to speak at the meeting of a business association in a small city in New England. He addressed the state official who has just introduced himself, and let loose a stream of angry accusations. The government had caused him misery with its crazy environmental rules. He was being driven out of business by a government that couldn't care less.

The whole room was filled with tension. The government official, new to his job, new to talking to business leaders, not sure quite how to respond, took off his jacket, as if he had just received a blast of terrific heat. The room burst into laughter. Even the angry company president laughed. The anger against government had flared up and been let out. But directing it at this poor bureaucrat, who hadn't yet said a thing, was silly, and everyone could see that.

From that point on, what the government official had to say was well received. The meeting ended with a warm thank you from the company president, as well as everyone else. The government official was me, and I had not expected this. How is it that when I was sent to the front lines of the war against pollution, a room full of polluters, that I ended up reporting back that there was peace and celebration? How is it that this wasn't just a moment, but that I was then invited to visit the companies, and we began relationships that lasted many years? I was not prepared to learn how well environmentalism and commercial capitalism could work together, how warmly an environmental official could be received by industry. I fear much of the public does not know this story, and it is not just my experience, but that of many officials in programs like mine.

It's not so easy to explain how this became a common phenomenon in Massachusetts and many other states, but one way to begin is to tell another story that began a few years later. The Churchill Coatings Corporation was experiencing such severe regulatory troubles in Massachusetts that it contemplated going out of business. It had a nice business pre-coating wood products—so successful that "pre-staining" had become the industry standard. But the air emissions of the sprays were volatile organic chemicals (VOCs), putting the company seriously out of compliance with new air pollution regulations. The company couldn't figure out how to continue operating and the Massachusetts Department of Environmental Protection (DEP) was pressuring them. By 2000, DEP's mandated reductions were "catastrophic" for the company's business. Fast-forward to 2007—and the company is now producing a new product called "ZeVo," for "Zero VOCs," and it is selling really well. How did that happen instead of them being forced out of business?

This outcome, like that of the first story, was the result of the application of a philosophy, a strategy, an approach, a way of thinking, a new way of doing business, and a new way for government to interact with business. It was once called source reduction (reducing the problem of pollution at the source), but became pollution prevention (P2).[1] It brought with it a new relationship between the government and the people, one in which the approach of offering assistance led the way (with enforcement backing it up if it failed). The message presented to the business leaders that day in 1990 was that the state had created a new agency—whose job was to try to help companies find ways to avoid creating pollution. The agency, where I was the first employee, was called the Office of Technical Assistance (OTA). One of the things I was able to say that day at the meeting of companies was that I would only visit a factory if invited, and unless I saw an imminent threat, I would not report anything. My advice to the company would be confidential, and this was required by law.[2] It was this message—not just "we're from the government and we're here to help," which is not believed—but "we're not going to tell on you, unless we have to," that worked. I was not going to knock on their door and say, let me in, I'm the law, and fine them for breaking some rule. I was asking them to invite me, telling them I was harmless, and offering help. This was different, and they liked it. There was relief that we weren't going to have a pitched battle with threats of fines if they should get caught. Nothing changed the fact

1. After 1989, it was called toxics use reduction in Massachusetts, but P2 has also been widely used there, as it embraces reducing pollution from substances not identified as "toxic." "Resource conservation" is also used to refer to saving other materials, energy, and water.

2. Credit here goes to Michael Brown, then head of the Office of Safe Waste Management (predecessor to OTA) who championed suggestions from myself and attorney Lauren Sloat that the confidentiality provisions under which we then operated needed to be incorporated in statute, and not just be office policy.

that another agency, DEP, was still going to inspect them, but I had told them I wasn't allowed to talk to DEP about them without their permission. If I did that, I'd be the one breaking the law. This was impressive to the businesspeople.

There are many stories about how well this approach worked, but the Churchill story is a particularly important one, partly because it is so well told by the company itself, but also because it has gone completely unrecognized. This is typical of the lack of attention to the many similar successes that the new approach of assistance brought about.

OTA's involvement with Churchill was the result of years of talking with DEP, to break down initial resistance. OTA Director Barbara Kelley and I continuously raised this idea and had to meet many objections, such as that companies would use OTA to play games and avoid proper enforcement. Over time, the enforcement agency grew to see that DEP could refuse to accept what a company said we told them (as an excuse for not doing the right thing) unless they were allowed verify with us what we had advised. They also saw how great it would be if we could solve a problem without the harsh procedures of enforcement, and moved to institutionalize the practice. In addition to successive Commissioners, Ed Kunce, Chief of Operations, saw this clearly.

When the company was told by DEP they would have to shut down if they couldn't stop emitting volatile chemicals, DEP urged the company to call OTA, and they did. OTA's Bill Griffin, who had extensive industrial experience, came back from his visit and told me that the emissions were unbelievable and the only way to stay in operation would be to totally change the chemicals they were using to pre-coat the shingles.

This then took a lot of discussion. OTA helped the company to perform trials of new chemical formulations, while we worked with DEP to hold off putting the company out of business. DEP was agreeable, not having any desire to put companies out of business, but it also could not tolerate the continued emissions. It was a difficult call and line to hold, and there was tension for several years in this regard. Only the company's diligent efforts, and OTA's representations to DEP that work was being seriously undertaken, staved off closure. Everyone's patience was tried and our honest reputation was at stake. In the end, it paid off bigger than anyone could have expected, but *no one at DEP ever knew that.*

In a 2007 presentation, Dennis Connelly, chief operating officer of the Churchill Coatings Corp., said that "We wouldn't be here if the Commonwealth of Mass. hadn't worked with us to find a solution. Mass. DEP and the

Office of Technical Assistance were instrumental and may even have helped to transform an industry."[3]

But the Churchill Company had moved out of state before the successful marketing arrival of their product, and I learned at a meeting of DEP officials some years afterward that no one involved in the matter knew how it turned out!

These stories should be a banner of hope in dark times. They should be better known. If they were, then the staff at OTA might not have been reduced from more than thirty, during the 1990s, to less than nine a decade later (a proposal to completely eliminate it generated dozens of letters of support from companies the office had helped).[4] What does it mean that our environmental efforts are so fraught with difficulties, and then when something works like a charm, no one pays attention? If things were the way they ought to be, OTA would have been expanded to meet the needs of the hundreds of thousands of businesses, municipalities, and individuals who need help knowing how to change resource use.

TWO KINDS OF BUSINESSES, TWO ARMS OF GOVERNANCE

There is a whole genre of feature stories that have been overlooked. They shed crucial light on how to win through this confusing time. There are many businesses that will transform. They are not committed to investments in poisoning the world. They are not willfully blind to their own impacts. We cannot just fight a defensive battle but must work with those ready to do business differently. The public interest is in changing the rules by which business operates so that we reduce the activities of those who harm public interests and nurture the development of businesses that benefit the public. This would be a shift from the current paradigm of assuming that promoting any business activity serves the public interest. The era of crafting rules to evolve responsibility involves the danger of overregulating but that is not a reason not to do what we need to do to control the impacts of our current economy.

---

3. At a conference titled "Cleaner Technology and Energy Efficiency: Structuring a Competitive Advantage," held April 5, in Boxborough, Massachusetts. Much credit, besides to the company, goes to OTA's John Raschko and the Toxics Use Reduction Institute, which provided substantial grant money for research trials, which included unsuccessful efforts with catalyzing solid coatings and electronic beam curing. In the end, the company succeeded with a combination of an old technology—linseed oil as the primer—and nonvolatile acrylic polymers, oven-cured, as the topcoat, to make the Zero VOC product.

4. It is possible to chalk up this blindness to the benefits of pollution prevention assistance to a focus on climate change that caused many environmental agencies to shift resources to promoting clean energy. This was similar to the early days of P2, when agencies shortchanged needed enforcement by shifting resources to prevention. It is a symptom of the failure of society at large to understand that we need to *expand* our investment in environmental health, not force shifts from one effort to another. We needed the expanded investment in clean energy, but we shouldn't have had to cut down a flourishing program to get it.

When we think of government as having two arms, and not just one, we can think of how the government presents itself in the relationship with the community. It can hold up a slapping hand and warn everyone not to get caught polluting. Or, it can extend a friendly hand, offer help in reducing impacts, and hold the other in reserve for punishment, if needed.

Governments in state after state, in municipal and county agencies, and at the federal EPA, began establishing the second arm—the helping hand—in the late 1980s. The United Nations and the Organization of Economic Cooperation and Development[5] picked up on the idea too. Rather than a big, brutal one-size-fits-all environmental governance was being personalized to the individual level, tailored to the group context. It was becoming more sophisticated and cooperative. It was moving from being reactive—acting when problems have mounted to intolerable levels—to what can be called anticipatory governance.

It was a philosophy, an approach, a method: *prevention.* And it primarily employed a particular means: *assistance*. It grew because it worked. Pollution prevention was started by private companies (3M's Pollution Prevention Pays program, which began in 1975)[6] and consultants,[7] and studied by public interest experts (Pollution Probe in Canada,[8] the Institute for Local Self Reliance,[9] INFORM Inc.,[10] and, most importantly, the federal Office of Technology

5. Shortly after we began publishing case studies of companies that had reduced toxics at the source, I received calls from many states, from OECD, from environmental policy-makers all around the world, asking for detailed information, and several environmental officials from European countries paid visits to our offices to interview us on how we were conducting our programs. In support of efforts around the world, the UN's Environment Programme published a definition of Cleaner Production in 1990: "The continuous application of an integrated environmental strategy to processes, products and services to increase efficiency and reduce risks to humans and the environment." UNEP now uses the term "Resource Efficient and Cleaner Production" to include resource conservation as a "key element of the transitions towards Green Industry and Green Economy," http://www.unep.fr/scp/cp/.

6. "Over the Last 40 Years, 3M Estimates that the Program Has Saved Nearly $1.9 Billion Dollars," https://www.pprc.org/wp-content/uploads/2015/09/P2-Retrospective_25-Year_Final-Draft_20151.pdf.

7. For example, Robert Pojasek, who later wrote with Cam Metcalf of Kentucky's Pollution Prevention program the EPA's 2001 *Organizational Guide to Pollution Prevention*, which, among other things, couched pollution prevention in terms of risk management and explained how to integrate it with quality, efficiency, and other programs, https://www.csu.edu/cerc/researchreports/documents/OrganizationalGuideToPollutionPrevention2003.pdf.

8. Notably, the 1982 *Profit from Pollution Prevention*, by Monica Campbell and William Glenn.

9. Their 1986 *Proven Profits from Pollution Prevention* is still online at https://cdn.ilsr.org/wp-content/uploads/1986/02/Proven-Profits-from-Pollution-Prevention.pdf. One of the authors, Donald Huisingh, later founded and edited the *Journal of Cleaner Production*.

10. A summary by one of the authors (David Sarokin) of INFORM's influential 1985 *Cutting Chemical Waste* found that chemical companies had reduced wastes using inexpensive measures at twelve of twenty-nine facilities studied (but that companies did not generally take this kind of action unless "their backs were to the wall," even though they could save substantial amounts of money, and none were methodically assessing how to reduce wastes; and that government regulations did not provide consistent incentives for companies to reduce wastes), https://p2infohouse.org/ref/28/27060.pdf.

Assessment),[11] and federal[12] and state agencies (North Carolina's Pollution Prevention Pays program, established in 1983). By the late 1980s, EPA was moving to institutionalize the more intelligent approach, and in 1990, Congress passed the Pollution Prevention Act, which found, in addition to the fact that immense amounts of wastes could be avoided, that:

> (3) The opportunities for source reduction are often not realized because existing regulations, and the industrial resources they require for compliance, focus upon treatment and disposal, rather than source reduction; existing regulations do not emphasize multi-media management of pollution; and businesses need information and technical assistance to overcome institutional barriers to the adoption of source reduction practices.
>
> (4) Source reduction is fundamentally different and more desirable than waste management and pollution control. The Environmental Protection Agency needs to address the historical lack of attention to source reduction.[13]

Because pollution prevention (P2) involves changes in operations, it is difficult if not impossible for the government to know what each business needs to do. For that reason, assistance, rather than the traditional approach of "command and control," prescribing specific actions by regulated entities, was the primary means by which the government could influence change. P2 regulation, when implemented, tends to be flexible in nature, imposing requirements that leave a great deal of discretion up to individual facilities. This is fine if we *don't replace, but supplement* enforcement, with assistance. Assistance will fail to reach many but must be used where it can work.

THE HELPING HAND HAS BEEN EXTRAORDINARILY EFFECTIVE

By 1990, every state in the United States had a P2 program. By 2003, it was indisputable that they had moved mountains. A compilation of results from only half of those programs, including only results that had sufficient documentation, indicated the national effort had reduced hundreds of billions of

---

11. The 1986 report *Serious Reduction of Hazardous Wastes*, by Joel Hirschhorn and Kirsten Oldenburg, found that prevention held huge promise but got little attention, and that "Current pollution control methods often do little more than move waste around" (4), https://www.princeton.edu/~ota/disk2/1986/8625/8625.PDF.

12. Such as the 1988 "Waste Minimization Opportunity Assessment Manual," by the Hazardous Waste Engineering Research Laboratory at EPA's Office of Research and Development (project lead, Harry Freeman), https://p2infohouse.org/ref/43/42046.pdf.

13. US Code Title 42, Chapter 133, Section 13101: https://www.epa.gov/p2/pollution-prevention-act-1990#pol.

pounds of pollution during the 1990s, with savings far exceeding the investment in P2.[14]

P2 simply makes sense. It makes sense to try to avoid creating something like hazardous waste if you have other options. *It makes sense that companies should look for those options, and if they are not looking for them, that government should act to get them to do so.*

But the use of assistance—extending the hand of friendship—presents a confusing picture when the battle over EPA is front page. It can look like the enforcement agencies are going soft, giving in.

Also complicating the picture is how stories about businesses stepping forward and voluntarily reducing their impacts can lead to complacency. The tension between government and industry can come to seem like the problem. If business is doing the right thing on their own, why do we have big bad government bothering them? And when government is not looking like a hateful giant, political opportunists seeking to ride resentment to power don't see any profit in telling the story that it is doing a good job. These are some reasons why the public seems not to have grasped that we had a Golden Age.

A SOMEWHAT ROCKY START

It is a legitimate concern that working in partnership with business can compromise the ability of an environmental agency to enforce. Shortly after the program was launched, I was asked to participate in a debate with Joel Blumstein, an attorney in the enforcement division at EPA's Region One office. The assistance effort was generating some fierce opposition from enforcement people, who said we would weaken the strength of their efforts. There had to be a scary deterrent message, not one that says the government is somebody you don't worry about. Part of the problem was that funds were being taken away from enforcement to finance the new assistance efforts. It was one of my first experiences of being brought in as one side of a hot debate and finding how things can cool down immediately by enlarging the context. I said at that meeting that the issue was not one or the other but how and when you use each and how they coordinate. This met with much agreement. I was asked to speak at many more such events, not just in Massachusetts but in other states as well. Everywhere I went in those early years there was anger among enforcement officials about the new emphasis on assistance. In many cases, that anger was justified, because the new emphasis was sometimes

14. "During the period 1998 to 2000, 13 P2 programs with a total average budget of $1.9 million annually reported total cost savings equal to $404 million," the National Pollution Prevention Roundtable, *An Ounce of Pollution Prevention is Worth Over 167 Billion Pounds of Cure*, https://www.epa.gov/p2/pollution-prevention-act-1990#pol.

articulated by disparaging enforcement's achievements. As a representative of the new assistance effort who praised enforcement, I was welcomed. It was actually easy to calm the roiling waters, and I was often amazed that no one had done it before. My theory is that those in high positions who launched assistance efforts for good reasons, did not have sufficient contact with those below them to know how people were feeling. I was lucky to get to see both levels, as a representative of the new approach, and I didn't get the impression that any leader was anti-enforcement, but employees often thought so and burned with anger. I saw it as a product of rigid hierarchical functioning and something that could have been remedied through more personal contact and honest sharing of thought.

The problems arise if you don't manage the effort. By setting protocols for how assistance and enforcement can coordinate, confusion and compromise can be avoided. In those first years, I spent a lot of time talking to enforcement agencies, working out agreements about how we would work together. Generally, the agencies were convinced to refer companies to us, and when they didn't need to take immediate enforcement action, they would see what we could do first. They eventually learned that we needed to maintain confidentiality, or companies would never let us on-site, and that they could trust us to inform companies of the law (assistance providers received training in the law, and worked hard to articulate what was required in ways that DEP and EPA approved of the language we used).[15]

The enforcement agencies also heard from us how the companies we visited were, generally, extremely unlikely to ever pick up a phone and call the enforcement agency to ask a question—most were convinced that this would trigger an inspection—but that they had lots of questions for me, because I was safe and the rules were very confusing. This was not always easy for the enforcement officials to hear. It took a few years for some of the hard-core enforcement officials to trust the new approach,[16] but eventually they did. It also took some years for a good portion of industry to understand that we were not in fact telling on them, and some have never come around.[17]

---

15. At this time many P2 programs were refraining from providing compliance assistance because of the concerns of the enforcement agencies that they would not do so correctly.

16. One circulated a memo calling OTA a "Black Hole" into which information disappeared. In response to this, we began sending weekly reports to DEP, which provided generic information about our work, withholding any information that could allow the enforcement officials to know which companies we were talking about, which was of immense help in creating a cooperative relationship with DEP. Barbara Kelley, director of OTA, instituted regular "information exchanges" between OTA and DEP to build mutual trust and to learn from each other.

17. On occasion, by mere coincidence, an inspector would visit a company after we did, and sometimes those companies believed we had told on them. When this happened, it took a great deal of outreach—going to meetings and showing people that the law prevented us from doing that—to repair the damage.

A LARGELY UNNOTICED GOLDEN AGE OF ENVIRONMENTAL POLICY

While this important advance in regulatory methods occurred, only a few noticed. A publication called *Merchandise Mart* ran a column by me. A Worcester, Massachusetts, paper featured stories about how we had helped local companies, calling us "Sweet OTA." But neither these nor similar successful policy innovations in state after state seem to have much impacted the public consciousness, which seems to remain firmly fixed on the idea that environmental agencies are frightening and cost jobs. Too much of the public still thinks government interference with business operations can't be anything but inefficient. Figuring out why the news that government can work hasn't penetrated would be useful.

UNREASONABLE FAITH IN THE IDEA THAT BUSINESSES REDUCE ALL COSTS

I will never forget the meeting—early in the program—at the Massachusetts Institute of Technology, to which I had been invited by environmental policy pioneers John Ehrenfeld and Jennifer Nash, at which I described how OTA worked. I said that we were going into companies and finding ways to reduce wastes and save lots of money. I had dozens of examples to point to, but an industry consultant got angry. He actually yelled at me, saying that I was insulting business leaders and clearly didn't know what I was talking about. If there was money to be found, these people would have already found it, he said. They are "expert money-hunters." I was a bit shocked by the intensity of the attack and did not have an immediate comeback except to say that I was not lying about the stories I was telling.

As I went home, the right response occurred to me. These people may be expert money-hunters, but they are *just not hunting in this part of the forest.* My experience had shown me that. When I made one of my first site visits, to a company that compressed gases into liquids, I was told many complex things that I did not understand about the physical process. There was a pool of oil on the floor under the machine, just staring us all in the face, but it seemed that only I saw it. I pointed to the oil on the floor and asked, "why is that oil on the floor?" I didn't need to say any more. They suddenly noticed that there was something wrong and went scurrying around trying to answer that question. Finally, they figured out that the person who maintained the machine unhooked hoses with hydraulic fluid and let them drip on the floor. It also turned out that this had been routine and the company was regularly cleaning up the mess with an absorbent and paying for the oil-drenched waste to be managed. The pool of oil on the floor had been ignored. It was no longer part of the production process, and was no longer important.[18]

Or when we visited a Fortune 500 company at which a jeering manager mocked the idea that we outsiders could find any ways to make the process better, and I pointed to an immense container—with about a third of it filled with oil—and asked why it was there. I was told that it was the oil in which a huge housing had been packed, and then when I asked what would happen to it, no one knew. After scurrying, it was learned that the container would be tipped and the oil poured into a drain. When I asked where the drain went, we were taken to a place called "The Ecology Pit." It collected such wastes from all around the plant. When I asked what happened to the waste in the Ecology Pit, there was more scurrying, and when the managers learned that the company was paying *30 thousand dollars a month* to pump the pit out and send that oil waste off, I didn't need to say anything. My questions about why the company couldn't reuse the oil for packing something else (it was perfectly clean) or whether it could use shrink-wrap plastic to protect the housing, instead of oil, were now received with respect. But it was amazing how no one had noticed the problem or that it was costing them $360,000 a year. *These companies needed a fresh pair of eyes to see the things they were not seeing.*

TRUST TAKES TIME TO BUILD

The first pilot study that got people from the Massachusetts government into facilities using dangerous materials was a project with T. R. Lakshmanan of Boston University, whose students went into metal plating companies in Southeast Massachusetts and performed measurements showing excessive water use and made recommendations for equipment that can recover metals from process water, so that they would not have to be discharged. The success of the project in getting on-site—to being invited by willing company owners—was due to support from the Southeast Regional Planning and Development District, where a man named Bill Napolitano, who knew all the local companies, helped develop the trust relationship necessary for assistance programs to work. Later, when I reached out to try to get on-site with printers (in previous years a large source of air emissions from alcohol-based fountain solutions), I succeeded because a man at the association Printing Industries

18. This principle, that businesses ignore aspects of operations not related to production, is the reason that the Resource Conservation and Recovery Act, our federal hazardous waste law, requires that generators ship their wastes to facilities licensed to manage them within set times, because if not required to do that, they might just pile them up until they become unmanageable.

of New England, named George Frantz, created that trust for me, vouching publicly for me at one of their annual meetings.[19] When Lee Dane, who came to the state agency from MIT's Sloan School, took me to my first meeting of plating companies, I saw Bill Napolitano tell the companies there that they could trust us. That we were good people. This is the same thing George told the printers years later. Think about this—it means that for the program to work we had to first counter the idea that the government people were bad. Lee's opening *joke*—her charming the business owners was another key factor in the agency being invited on-site—was that she was from the government and was there to help them. People were at that time treating this as a terrifying concept.[20] Of course government can sometimes be awful. That doesn't change that it can sometimes be great.

I remember at that meeting how simply explaining the environmental rules—once again, we were invited because the rules were confusing, and we were going to clarify things—got one guy so mad he stormed out, angry. Someone told me that he had expected to inherit his family's business but now he was under the gun to pay for cleaning up land at the company that was contaminated, and he didn't know how he would survive.

At meetings like this we slowly built trust, a process that took years. To explain how that happened would take describing many persistent efforts, to help, to show how we could help, and overcome the fear and suspicion that prevented us from helping. One remarkable moment came when Tim Greiner, a young engineer just out of the Rochester Institute of Technology, visited a shop that performed a "phosphating" process, as part of the Blackstone Project.

THE FUTILITY OF MOVING POLLUTION AROUND

Manik (Nikki) Roy, who went on to work at EPA headquarters and more, was then at the DEP, and he was working with Lee and our boss, Michael Brown,[21] chief of our nonenforcement Office of Safe Waste Management in the Department of Environmental Management,[22] on a plan to address

19. George later came to work at OTA, where he led successful efforts to establish partnerships with small business industry sectors, such as auto shops, printers, and dry cleaners, and then moved on to EPA.

20. President Ronald Reagan had not helped matters when he said—many times—that the nine most terrifying words in the English language are "I'm from the government and I'm here to help," https://www.reagan.com/ronald-reagan-on-big-government-most-terrifying-words.

21. Brown went on help Patagonia implement source reduction and to found Brown and Wilmanns consulting.

22. OSWM became OTA in 1989 when the Massachusetts Toxics Use Reduction Act (Chapter 21I of the Mass. General Laws) was passed.

the "cross-media shifting" of pollution. This shifting occurs because of the way environmental protection is divided into "silos" of media—air, water, and waste. First you send out your air inspector, and she finds excessive emissions and orders the company to treat them, and the company puts on a scrubber, which makes a wastewater. Then the company gets visited by a water inspector who says you have excessive discharges and orders the company to treat them, and the company buys a big wastewater treatment process and hires an operator qualified to operate it and makes a sludge they have to dispose of. And then the company is visited by a waste inspector who says you can't dump that sludge, and now it has to pay a lot of money to dispose of their solid waste. How much more efficient would it be to do them all at once? Then maybe that will get them to think about the *source* of the problem, the use of the substances that cause pollution.

I asked why we couldn't also include referrals for technical assistance? Why couldn't the enforcement people tell the companies about us assistance people? And suggest that they see if we might not be able to help? DEP agreed and we wrote it up and got an EPA grant for what became known as the Blackstone Project, because we included someone from the sewage plant discharging to the Blackstone River in Central Massachusetts.[23]

Tim Greiner was the first full-time employee of OTA who had a technical degree (my degree was in law).[24] To understand how powerful it was when Tim visited the phosphating shop, it is important to know that for two years I had been told by many people that the assistance program was not a practical idea. Some said that companies would not want to accept my offer of technical assistance (and during that time only a few were doing so). They would not want to risk someone from government finding a violation and telling on them, even though I was promising confidentiality and even though I could provide a legal argument that I could not tell on them. Companies would also not wish to share their successes, I was told, because they would be competitive advantages. If it was true as I said that hazardous waste generation was a big cost in terms of time and liability and management costs, then they would produce more cheaply, and why would they want their competitors to know

23. This was Paul Richard, who was named director of OTA by Gina McCarthy when she was coordinator of Toxics Use Reduction programs in Massachusetts. I had visited the sewage treatment plant while his pretreatment program (which works with companies discharging to the plant) was being examined by EPA, and witnessed his detailed answers to every question.

24. OTA was also greatly aided at this time by Joe Paluzzi, an experienced engineer who remarkably started as an intern because of his belief in the approach, and Mitch Kennedy, fresh out of college. Our director, Barbara Kelley, had headed up efforts to foster household hazardous waste collections, and constantly emphasized the need to get out and meet people to form collaborative initiatives. My gratitude for her support and guidance cannot be expressed, and her wisdom was universally respected. Though she kept a sign on her door that warned visitors against whining, she demonstrated the value of listening and empathy to all.

how to do that too? And finally, these skeptics said, these companies have spent years refining their processes. How was I or anyone from the outside going to come up with reasonable ideas for changing their processes? Their products are made just so for a reason! I knew enough to doubt this but was completely uncertain of our chances of getting results.

The company Tim visited had been having constant violations of wastewater limits. They were making wrenches that would be dipped into a bath of a phosphate solution and then rinsed off. The discharged wastewaters were exceeding legal limits. The company kept getting fines and it was costing them more and more, and they didn't know what to do. This is, of course, what makes people hate the environmental agencies. They are trying to make money and they see this unexpected cost and this is just what they have to do to make their product and the fine comes suddenly—it is very frustrating. Tim walked into the facility and saw that there was an empty tank in the corner. He said, if you bring that over here and fill it with water, and then dip into that first, before the rinsing (which goes down the drain), you will get most of the phosphate solution off the wrench, and it will remain in this tank (it's called a "still rinse," or "dead rinse"). At the end of the day, Tim explained, you pour some of it into the phosphating tank, which is hot, and it will make up for evaporative losses and recover the materials for continued use.

The company was essentially instantly back in compliance. Word spread from this event, the company president was so grateful he told his friends, and Tim was invited to several other companies.[25] We *could* find answers. And most amazingly, companies were not just willing but many were eager to share the story about what they'd done. We wrote dozens of case studies, hosted scores of meetings where companies shared information, even organized demonstration site events where companies invited their competitors to come and see what they'd done at their facilities.[26]

"We're all in the same boat," said Dale Broadbent, the president of Reliable Plating, when asked why he was willing to share his story. He had allowed the Boston University students to come in and they had calculated

25. Tim continues to be invited to companies to help them reduce environmental impacts, as a founder and principle of the consulting firm Pure Strategies.

26. One event, the brainchild of Tim Greiner and Mitch Kennedy, was the "Solvents Bazaar," which invited companies to bring dirty parts to be cleaned by companies selling safer alternative cleaners. Ed Surette came from M/A-Com, a microwave equipment manufacturer using Freon to clean. Surette was inspired by the event to launch an effort to evaluate alternative cleaners, saying he learned from OTA that you have to look closely at how something cleans, and that by using a "black light" device, he determined that a low volatility alcohol actually cleaned better than the Freon. The company switched just as the tax on ozone-depleting chemicals was instituted and saved an estimated $8 million dollars as a result. Surette then took what he learned to M/A-Com facilities around the world. See The Merrimack Project: Building Partnerships to Prevent Pollution in a Watershed, Commonwealth of MA, Executive Office of Environmental Affairs, US EPA Industrial Pollution Prevention Project Grant X1001588-01-0, June 1995.

the money he spent on managing hazardous waste that could be saved if he stopped using trichloroethylene (TCE) to strip the grease off the metal he was plating. He told me that he used to buy his metal parts from a nearby company, but that when it was so much cheaper to get them made in Asia, he did that. But the parts from Asia came coated in grease, to protect them from tarnishing during transit. He needed the TCE, but its use created a TCE/grease waste that cost a lot to get rid of. The hazardous waste law, RCRA,[27] now made generators of such wastes responsible for them from the cradle to the grave, and the Superfund law, CERCLA,[28] made paying for that grave very expensive, if it didn't properly contain the wastes. Little companies like Broadbent's now even had to research the people who took their wastes, and the places where they sent it, to make sure that they were properly handled, or they might be subject to huge cleanup costs.

The BU students showed Broadbent he was saving nothing by buying cheap products from Asia, and he stopped. He bought from the nearby company again, he told me, because their parts didn't have to be coated in grease. He got them fresh off the press. He was delighted not to be using TCE. In my first visit to a plating shop, I was told by a consultant that the longtime workers often had slurred speech, because the solvents like TCE had stripped the fatty myelin from their nerves, just as it stripped the oils from metal parts.

Reliable Plating's was one of the first stories I was allowed to tell others, and when I asked him why he was letting me do that, wasn't it a competitive advantage to keep this to himself, he said that he was proud he was buying from a local company again, but more than that, he said, there had been a spill of TCE in the news lately. He said it was a company like his, and after the spill, he saw the name of his company in the newspaper on a list of TCE-using companies. He had an important self-interest in keeping the companies like his out of trouble like that.

I met many people in business who were motivated by heartfelt connection to others. My experience with providing assistance certainly showed me many cut corners and many violations, and I didn't even see that which companies kept to themselves, but to an environmentalist on the other side of the pollution controversy, it was surprising and encouraging to see many voluntary responsible actions. I began to realize that the polluting industries around me were filled with people who would love to not have the problems that vex the rest of us, and many were led by people who cared about their community.

---

27. Commonly referred to as "REK RA," the Resource Conservation and Recovery Act.

28. "SERK LA," the Comprehensive Environmental Responsibility, Compensation, and Liability Act.

FAIRER TREATMENT OF THOSE WITH GOODWILL

This was an important thing for me to learn. As I got to be friends with so many in the industrial sector, I realized I was meeting many good people. I worked to promote trust between them and the enforcement agencies. I saw so much unnecessary suspicion and fear of each other. This could, however, be worked through. I saw significant changes take place as a result of greater focus on recognizing and cultivating responsible behavior, as opposed to only punishing the irresponsible.

The enforcement mechanisms of environmental agencies have matured from the more primitive early days when the relationship was cruder and gave rise to bad feeling in the regulated community and in the enforcement community. Enforcement now—to some extent—takes goodwill into account, refers to assistance, and seeks settlements that restore both environments and reputations.

Enforcement processes are now standardized, so that people are treated the same.[29] Calculating a fine should now take into account how someone responds to being caught, crafting a better path for those willing to own up and do the right thing. An entire methodology has been developed for how companies can take steps to make amends. Agencies developed outreach efforts, guidance, grants, partnerships and targeted efforts to help whole sectors. Some have worked hard to help companies who want to know what to do and will sometimes even suspend enforcement (forbearance, or amnesty) if they think people just don't understand that a law exists and a new education campaign is necessary before it's fair to fine anyone for violations. EPA and many state agencies began giving companies credit for having an environmental management system in place, distinguishing between a violation that is a mistake and a violation that is the result of a wholesale ignoring of responsibility. All of these and other initiatives show that agencies have grown and changed their way of relating to the companies they regulate, in order to be fairer to them. But there is so much more to do.

THE DATA PROVES THE PREMISE

The assistance programs worked and accomplished reductions in toxics use far greater than hoped. The Massachusetts Toxics Use Reduction Act, which established the assistance effort, also requires that companies do TUR plans—that they examine their opportunities to reduce the toxics they use.

29. This is not a statement that standardization has been universally achieved, but that standards are now well developed for use.

The data from the companies covered under TURA showed that the majority of covered companies found ways to reduce toxic chemical use, without the government telling them to do anything in particular, just by being told by the government to look into it themselves. The act is essentially *the idea that the companies owe this attempt to reduce their own impact—to try to reduce toxics use—to the community.*

Companies also self-reported that these actions saved them a lot of money, totaling more than what they said compliance cost them. The first program assessment, conducted in 1997 by Abt Associates, found substantial increases in attention to toxics, significant reductions in toxics use, waste, and releases, and savings far exceeding costs.[30] TURA requires users of large quantities of toxics to report on what they use, and analysis has shown that:

> from 1993 to 2002, chemical input by all companies combined was approximately 500 million pounds less than what was expected. The study showed that for every company covered by TURA requirements that did not improve performance, there were nearly four companies that did improve. During the period examined, about one-third of all chemicals "dropped out" of the program: that is, they came to be used in amounts below the threshold for reporting.[31]

Although some have pointed out that reductions have never been as dramatic since the first years of the program, that is to be expected as companies learn to operate in a new way, and the "low-hanging fruit" of easily implemented opportunities is harvested, and the more complex and difficult problems predominate. That the rate of reductions decreases is evidence that companies are learning. Ironically, opponents of the law point to what is now more common good practice as evidence that the law is no longer needed.

But as people in the field know, "low-hanging fruit grows back." This also happens more frequently the more we forget the lessons we have learned. That P2 is largely forgotten, programs like OTA have diminished instead of expanded, and we shifted attention to climate change without retaining a focus on de-poisoning or explicitly marrying the two to best synergistic effect, has exacerbated this condition.

Another reason for increasing, and not decreasing, attention to P2 is the fact that some toxics use is not easy to eliminate. This is an argument for increased research and development, and increasing both technical and financial assistance.

---

30. https://www.turi.org/Our_Work/Policy/Toxics_Use_Reduction_Act/Program_Assessments.

31. Executive Summary of *The Effect of Providing Onsite Technical Assistance for Toxics Use Reduction*, 2006 (principal author, myself), was the result of a two-year analysis of TURA data, vetted by data experts from Harvard, EPA, MIT, and elsewhere, using simple math so that it could be explained and checked by anyone (statistical methods were only used to confirm results): https://www.mass.gov/files/ota_effectiveness_study_final_2006.pdf.

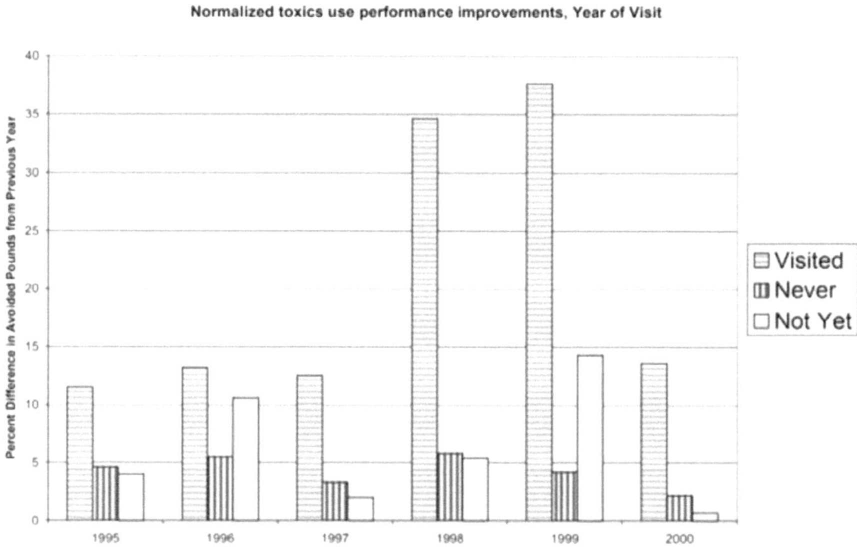

Figure 3.1. Three-Year Comparison—Chemical Use Reductions, Average changes in punds of toxics reduced in terms of percentages of use, comparing performance in the year of the visit to the year before.

We learned from TURA that a flexible regulation that simply asserts a responsibility and then requires reporting that flows from that responsibility works. The fact that so many companies ended up better off—according to data showing reduced toxics use, reduced emissions, and reduced waste—and also according to their own reports on money saved and the choices they made to invest in prevention—should lead to public support for more such laws. Use of third-party verification should be included in legislation to expand on these lessons. Each plan must be certified by a qualified verifier.[32] It also shows that the effect of *chemical input reporting* is important—companies having to track their chemical use tend to notice when the chemicals are being lost, or used inefficiently. These are tools of governance—that provide shifts in attention—that could be more widely applied.

Most gratifying for those of us in the assistance profession was that the data showed that providing the kind of one-on-one assistance of the OTA,

32. In Massachusetts, this is done by a "Toxics Use Reduction Planner." TURPs formed an association and have created and upheld a preventive culture in the consulting community.

that combines compliance and technical help, made the law work doubly or triply well.[33]

OTA's visits were associated with reductions in hundreds of millions of pounds of toxics that companies implemented voluntarily (figure 3.1).[34]

It is important to note that companies can comply with TURA's planning and reporting and *not reduce one pound of toxics use, and they won't be penalized* unless their plans and reports are fraudulent. These reductions were actions taken because companies thought they made sense.

THE REAL STORY IS MUCH LARGER THAN WHAT THE DATA SHOWS

Massachusetts is no lonely example. Hours after TURA was passed, Oregon passed a pollution prevention law. The State Roundtable on Waste Reduction programs, founded by North Carolina, became the National Pollution Prevention Roundtable to serve the programs operating in every state, and to work with EPA's P2 programs and hold regular national conferences;[35] the Pacific Northwest Pollution Prevention Resource Center was founded in 1991, the Great Lakes Regional P2 Roundtable in 1994, and the Northeast Waste Management Officials Association was at work coordinating prevention efforts at the regional level since the beginning of the movement. There were innovative programs in so many states that to mention California, Texas, Ohio, Vermont, Illinois, Oregon, Washington, Connecticut, New York, and Kentucky seems unfair to others. Because only two states (Massachusetts and New Jersey) require reporting of chemical input, the successes of these programs can easily remain invisible. In states that have small staffs doing P2 work, there may be little ability to do the research to document results.

33. *Effect of Providing Onsite Technical Assistance*, 22. The companies visited by OTA did much better than those not yet or never visited, by virtually every way you could measure performance. The one exception was that some companies perform very well on their own—they don't ask for visits because they don't need help. But there were not as many of those as you would wish. See also: "Does Providing Technical Assistance for Toxics Use Reduction Really Work?," *Journal of Cleaner Production*, September 2008, https://www.sciencedirect.com/science/article/pii/S0959652607001990.

34. I was not permitted to include this estimate in the report. In a rare interference with my work, I was instructed to tone down the estimate to the most conservative (still large, at 63 million pounds), on the grounds that my estimate would not be believed. But a statistical analysis performed by a Boston University colleague confirmed the larger estimate I calculated, and I hold to it, especially since the data only concerned the data reported associated with visits, and much advice was given over the phone, at meetings, conferences, and through guidance, the results of which we never saw, and the study only covered a six-year period, while OTA has been at work for more than three decades now. In figure 3.1, companies visited by OTA performed markedly better than those who were never visited, or had not yet been visited by the year noted.

35. Funding for these and other projects have dried up. This is one reason I'm writing this book.

They have documented impressive savings when measurements have been made. But the real savings are likely far larger than we can document, for every story that gets told, there are many that are not. OTA and these other programs visited and talked to many companies and never found out what happened as a result. Identifying the results, key to convincing administrators and legislators and the public that your program deserves continued funding or expansion, is hard work.

For example, around 2005 I was engaged in trying to document what had happened with companies that OTA staff had visited. We knew that Berkshire Manufacturing, in the western part of the state, had eliminated a toxic chemical, so I drove out there and sat with the engineer who did the work, who expressed gratitude for the help and a desire for others to know about what had been achieved. He was proud of it, and it had saved the company tens of thousands of dollars a year, for which he was very grateful. He had spent the time to gather up records for me. This was the work most companies did not do. They took our advice and got reductions and that's good, and it's understandable that they didn't have the time to sit down and calculate their savings. That was our issue, not theirs. But he had done it. He showed me hazardous waste cost reports and air permits and we totaled up his annual savings, while I asked him questions about why and how he did this. The company had not given a thought to reducing the chemical—which did its job perfectly well—until it was tightly regulated because it is a serious air pollutant. I kept asking questions about what happened as a result of this project. After more than an hour of talking about avoided waste management costs, he sat back and said, you know, come to think about it, we saved a lot more than a few tens of thousands of dollars in avoided waste management costs. He said, "we saved more like four million dollars." He said that the company's major customer, the one worth four million in profits every year, decided to stop ordering from suppliers using the chemical, and had Berkshire not made the switch, they would have lost the business. The savings in waste management they didn't have to do were small in comparison, and he hadn't thought of that until we'd been talking for over an hour.[36] There are many stories like this, all over the country, all over the world, that have not been told, or have not been heeded.[37]

---

36. Unfortunately, website management efforts by the state have resulted in the removal of many old case studies such as this one from the online collection of case studies. But they will have been archived along with other work taxpayers paid for, of lasting value. Some, too, remain in my basement and probably in the basements of other retired environmental officials proud of their work.

37. For just two concise sets of examples, see https://www.pprc.org/wp-content/uploads/2015/09/P2-Retrospective_25-Year_Final-Draft_20151.pdf, or https://www.turi.org/Our_Work/Policy/Toxics_Use_Reduction_Act/Companies_Reduce_Toxics_Save_Money/25th_Anniversary_Leaders.

TWO NEW, BARELY NOTICED PARADIGMS TO WHICH WE OUGHT TO SHIFT

It makes sense to think about this evolutionary period as one of two revolutions, two paradigm shifts; one from the dominance of the focus on managing wastes after their creation to recognition of the primacy of the objective of preventing their creation; and one from the dominance of the enforcement mode to one of a facilitating relationship. If the story of the evolution of environmental governance advancing in these two simultaneous ways were better known, it might be easier to generate support for reviving and expanding them. The Earth's biosystems desperately need humanity to transform itself so that destruction of life itself will cease. Pollution prevention is a proven method for better and more effective environmental protection. Why has it been resisted?

One obvious reason is that it reduces sales of product, and providers of toxic chemicals are financially affected. In Massachusetts, shortly after the source reduction programs began demonstrating the effectiveness of the approach, the Chemical Manufacturers Association (now the American Chemistry Council) funded the creation of the Massachusetts Chemistry and Technology Alliance (MCTA), which launched annual campaigns to repeal TURA, and lobbies against operation and expansion of the program. I spent much time during the early years of the program talking to people in many states who were working against chemical industry opposition to establish programs similar to the Toxics Use Reduction Act. These efforts were beat back, as was a national effort during the Clinton administration to build chemical use reporting into the Emergency Planning and Community Right to Know Act.[38] We in the program also spent a lot of time countering misinformation about the act, which claimed our successes were illusory and that we were harming companies. I used to distinguish between chemical industry people who questioned our successes with respectable arguments and those who outright lied.

---

38. Known at the time as "TRI (Toxics Release Inventory) Phase Three," an advanced notice of proposed rulemaking was published in the *Federal Register* on October 1, 1996 (see https://www.gao.gov/assets/230/224282.pdf for a report by the General Accounting Office). I remember being told by upper-level EPA officials that the plan had wide support and would "beyond a doubt" go through. In my opinion, exaggerations of the burden of chemical use reporting and vast underestimates of its value, promoted by the chemical industry, doomed the effort. Not enough people understood that chemical use reporting directs the attention of companies using toxics to where they are losing materials—that it helps them to be more efficient—and that the burden quickly vanishes to negligible once a company knows how to do it and has it in place. From there on, the benefits of information and efficiency accrue. The losers are the companies that sell toxics and the companies that continue to use toxics who must compete with cleaner alternatives.

The environmental field has developed new methods of governance that work, that need to be added to—not to replace but to work with—the traditional methods (often referred to as "command and control"). EPA and the state agencies fostered a tremendous amount of change that became new ways to do business, and have largely been identified by the public as things companies have done, which is largely as intended. The assistance programs succeeded when companies got credit for what they did. But this renders the roots of the change less visible.

Not a few entities, companies and institutions that boast about their environmental actions were started on their road—shamed, shocked into awareness—by environmental enforcement. And many are the companies who became greener because of their own commitment, but were prompted to start down that road by something a pollution prevention program did. The value of government action is not going to appear in easily accessible tables of numbers. The ecosystem of information, ideas, cultural orientation, and economic motivation must be understood for the relationship of cause and effect to be seen.

For example: What was the value of the "immortal etchant" story I arranged to be told, to the dozens of companies in Massachusetts using strong acids to etch metals, in dangerous facilities that had discharge and emission issues? On the letterhead of my office, with presidential candidate Michael Dukakis as governor, I reported that "As a result of information imparted to a key member of the New England chapter of the Photochemical Machining Institute, on October 27, 1990, in South Attleboro, Massachusetts, Dr. Lawrence David of IBM addressed an audience of institute members on the subject of extending the life of etchants." The members had doubted my suggestions for acid life extension would work. They were stunned to hear that Dr. David had two successful strategies to share.

One process involved regeneration of ferricyanide etchants with ozone. Bubbling ozone through the spent etchant reverses the oxidation-reduction reaction that occurs in the etchant process. In the past, IBM disposed of this spent etchant every twelve days. As a result of the successful regeneration of the etchant, it was continuously reused for several years. Because the process is stabilized, the performance of the etchant has improved and the rate of throughput increased by about half. Dr. David estimated the improvements, costing $100,000 to implement, saved IBM $250,000 per year.

The other process was one of precisely calibrating and controlling the pH every day. This also had extended the life of the chemical bath so that they were basically not having to dump it all the time. After a few months of going to business meetings and telling company heads that I had read about how people were learning to do source reduction—and asking if they'd ever thought about it and would they be willing to try—what a relief it was to have this expert do it, and what he had to say was so much more powerful. Dr. David said that because the etching is on a product that has already proceeded through many stages of production, the value of stabilizing the process was immense. Fewer rejects from the etching process resulted in estimated savings of millions of dollars, dwarfing the avoided disposal costs of $25,000 or the time savings of $42,000 per year. Dr. David commented:

> Turn back the clock thirty years—let's say you can put it (waste) directly into the Hudson River or Boston Harbor—you'll still be committing financial suicide [if you don't seize the opportunity for process improvement].

I got a call shortly after the talk from the owner of a small company using etchants, and he told me that instead of measuring pH, he had started measuring baumé, which is a unit of specific gravity. I had never heard of baumé before, but he said that didn't matter. The point was that he had started doing specific measurements, as Dr. David recommended, and he was maintaining the needed level precisely. This already had significantly reduced his waste. We got many visits as a result of that talk, and soon we had a demand we could only meet by having more meetings and conferences on acid waste reduction.

For years I would visit companies and find out that they had reduced their acid wastes, and I would always ask, What started you doing that? And they would say that they heard about it from someone who had done it. Sometimes they would credit our conferences, but usually they didn't know about where the idea came from, but the time of the change was never before the conferences.

Similarly, the work we did to bring awareness to the auto repair sector, of how to reduce use of and exposures to the many toxics that so many employees experience, resulted in changes that weren't later perceived as coming from our program. George Frantz—who had left the Printing Industries of New England to work with us—led the development of guidance for auto shops with a cleverly titled CRASH Course—Collision Repair Auto Shop Help. George Frantz, an intern named Stephen George, and I used EPA grant money (so much was done with EPA grant money when it flowed, and at very small expense) to work with companies in the field to identify the best practices that shops could implement. We worked hard to make this a product put together with the help of shops themselves, and with the approval of the enforcement agency. We bridged the two worlds and produced something shops could easily read and follow, and which the agencies would bless. We felt that we brought clarity and some relief, as well as increased safety, to this industry sector. For many years, if I talked to someone in the field, they would speak with appreciation of the CRASH Course, and tell me how it helped them change practice for the better. But that was a long time ago now, and the effort has not been renewed. It was more than ten years ago now that I had work done on my car, and the shop owner pulled out a worn and tattered copy of the document, and thanked me for it.[39] But that was the last time anyone remembered the CRASH Course. It seems that many have implemented the practices without knowing anything about OTA.

We reached out to Vocational Tech schools to keep their students safe, and just a couple of years ago I found myself giving an award to one,[40] for excellence in teaching green practices, and it was a school that I had visited many years earlier. Somewhere in a pile in my basement I have a copy of the detailed letter of recommendations I gave them at that time, on how they could green their operations and their teaching, to help their students work in that safer world they could help create. Yet when I gave the award, they didn't know who I was or what I had done to start them on the road they went down. Nor was it necessary for them to know. But it is necessary for taxpayers to know the value of such programs.

In a paper written by four state pollution prevention practitioners, a section entitled "Declining public sector support" observes:

> Through the 1990s, pollution prevention efforts were flourishing in both the private and public sectors. . . . By 1992 at least 25 states had passed pollution prevention legislation and by 1997 every state offered some form of pollution prevention assistance, over 30 operated information clearinghouses, and 30 states had some type of pollution prevention planning program.[41]

Since then, many pollution prevention efforts have been integrated into regulatory programs in government agencies. The goal of this was to make P2 an integral part of each agency's way of operating versus being in a separate program or office; in USEPA, for example, the Office of Pollution Prevention was moved from the Office of the Administrator to the Office of Prevention, Pesticides, and Toxic Substances (OPPT). The result often has been a loss of visibility, focus, and accountability for pollution prevention. Amongst businesses, some have the idea that they have implemented pollution prevention practices and it is time to move on to more pressing issues or greater

---

39. This project also pioneered the policy of recognizing good faith efforts, an advance that has been forgotten. See "Good Faith as a Fundamental Principle of Relational Environmental Governance," http://www.bu.edu/earth/files/2012/08/goodfaith.pdf.

40. The National Pollution Prevention Roundtable's Most Valuable P2 award, 2019, for P2 Project, Dan Capuano and Ken Stukonis of the Assabet Valley Regional Technical School, https://nppr.wildapricot.org/2019-MVP2-Awards.

41. *Pollution prevention 1997: A National Progress Report*, United States Environmental Protection Agency, Office of Pollution Prevention and Toxics.

opportunity. However, there are numerous pollution prevention opportunities in just about every business that have not been realized.[42]

The taxpayer paid for me to learn about things like acid recycling or solvent substitution, and get others to know about them. This was an investment, and it paid off. But an investment's value is not realized when it is discarded or retires without training a replacement. OTA and all the programs like it should have been expanded, not reduced. Yet when the Toxic Substances Control Act was being reauthorized, Republicans tried to preempt all state laws on toxics.[43]

The disinvestment in something that worked reflects a nationwide trend, also reflected in the elimination of the national OTA—the Office of Technology Assessment that published the seminal work on pollution prevention, *Serious Reduction of Hazardous Waste*. I once asked a knowledgeable policy person how Congress could ever have eliminated something useful like the Office of Technology Assessment, which only helped it to understand difficult issues, and he said that when there is pressure to cut government costs, people look for something they can cut. He said it may not mean anything more than that. To me, cutting something that makes you smarter confirms how dumb it is to just look for something you can cut. The conservative revolution that caused that to happen was simplistic. It flows from an assumption that the smaller government is, the better. Even Henry David Thoreau, who began his essay "Civil Disobedience" with the statement that he "heartily accepted" the motto "That government is best which governs least," clarified in that famous writing that what he really wanted was "at once a better government."

> Let every man make known what kind of government would command his respect, and that will be one step toward obtaining it.

We need to be sophisticated and understand the programs of our own government. It belongs to us. The taxpayer paid for programs that work. The taxpayer should know what a good investment in our future they are. The informed taxpayer should make known that he and she want government that works.

---

42. Gary Miller, Jeffrey Burke, Cindy McComas, and Kevin Dick, "Advancing Pollution Prevention and Cleaner Production: USA's Contribution," *Journal of Cleaner Production* 16 (2008): 665 015072. Miller worked for the Illinois program, McComas Minnesota, and Dick Nevada; and Burke was the director of the National Pollution Prevention Roundtable. The article describes the difficulty in obtaining data showing program success.

43. The standard argument for preempting state laws are that dealing with a patchwork quilt of many laws in different states is an economic burden. The question is not whether state toxic laws and programs discourage the purchasing of products that become hazardous waste, we assume they are or producers of such products would not be in opposition. The question is who bears the burden of having or not having toxics in our goods.

Meanwhile, the auto shops are facing new hazards. Cars are made with different materials, which behave differently. If you have more aluminum, then you can't use a torch the same way, and the dusts are quite flammable. You have an electrical system you need to understand.

The National Pollution Prevention Roundtable almost went bankrupt. In my second stint on the board, I watched years of budget reports where we were in the red. We are finally in the black once again, basically breaking even. There are new members. Pollution prevention programs still exist, but barely. The idea of P2 is no longer on everyone's lips, and the idea of voluntary programs and assistance seems a distraction in an age when environmental enforcement has been nearly destroyed. But it can come back. If we are aware of its history, more will call for it to be brought back.

The same thing that happened to P2 also happened to its big brother, energy efficiency, which Joseph Romm correctly but uniquely terms a form of pollution prevention. In 2007, he wrote about the Bush administration's proposal to zero out the Industries of the Future program of the Department of Energy:

> Pollution-prevention technologies take a variety of forms. Renewable energy prevents pollution in the production of electricity. Fuel cells offer the hope of preventing pollution in the transportation sector. Many other sectors of the economy have equally great prevention opportunities . . .
>
> . . . a DOE investment totaling about $1.1 billion in energy-efficient industrial technologies has yielded "approximately $2.5 billion in documented energy savings and net productivity gains, and the accumulation of these savings continues to grow at increasing rates." By 2000 these investments will be generating savings of about $10 billion a year. Very few other federal investments produce as great a societal return on taxpayers' dollars. . . .
>
> Funding for pollution prevention is the best way for the nation to avoid the need for costly environmental regulations. The government has a role in encouraging pollution prevention for several reasons. First, pollution-prevention technologies often benefit each of many companies only a little bit, so no one company has an incentive to spend the necessary money by itself. Second, prevention has many societal benefits: it reduces energy and other resource consumption and improves the environment, among other advantages. Third, and most important, pollution prevention and resource efficiency help companies to shift money from consuming energy and resources to investing in technology and capital equipment, thus creating jobs and economic growth. Indeed, a shift from consumption to investment may be the single most important transformation the US economy must undergo if we are to remain prosperous in the next century.[44]

---

44. Joseph Romm, "Pollution Prevention and Preparing for the Future," *Grist*, November 5, 2007, https://grist.org/article/mideast-oil-forever-part-v/. Romm is founder of the blog *Climate Progress* and was an acting assistant administrator at the DOE.

Romm noted enormous documented savings. "A 1993 analysis . . . found that a 10–20 percent reduction in waste by American industry would generate a cumulative increase of $1.94 trillion in the gross domestic product from 1996 to 2010. By 2010, the improvements would be generating two million new jobs, or roughly 1.5 percent of employment in that year."

Yet, Congress had already substantially cut these programs and now the administration was bent on eliminating them. Romm concluded: "Only a misbegotten ideology could conceive a blunder of such potentially historic proportions." The paradigm shift to prevention occurred and it worked. But then it was turned back. The Bush administration's well-documented fealty to the fossil-fuel and chemical industries is an available plausible explanation for why they did what they did, as well as for the actions of conservative members of Congress dependent on their campaign contributions and responsive to voters convinced by their propaganda. But it does not explain the failure of others to see the opportunities we lost, and to turn the ship of state back on the course it had set, toward more promising shores.

DEVELOPING ENVIRONMENTAL GOVERNANCE

The history of the success of the preventive approach, and of assistance coordinated with enforcement, should encourage us to expand, not contract, environmental governance. It should encourage us to ask for cooperation, to convene efforts to attain the common good. We can have a many-handed approach, which uses the full toolbox of strategies and mechanisms. We can give out grants and have rewards and preferences and structure incentives so that we encourage the responsible and discourage the recalcitrant.

It takes expanding environmental agency budgets, hiring people of various skills, cultivating an art of governance that is new to many people and many places. Assistance takes scientists, engineers, people who can figure out the law, people who figure out how to reach the target audience, people who can work with the conflicts that arise, and people who can understand what motivates others to take the desired action, to mention a few necessary parts.

Using the hand of assistance in conjunction with the hand of command and control implies that there's a head coordinating the two. With that image in mind, we can picture how the entire system can improve. When you have only the enforcement mode, businesses run and hide, and you only learn what you can learn when you inspect and find violations. The process of enforcement then focuses on those violations and not on the roads, not taken, that could have avoided them. But with the hand of assistance, you can get on-site and develop a relationship. Not only can you then explore alternatives with the regulated party, but you can also learn about how the requirements themselves don't always make sense.

A lot of undesirable activity occurs because people have not received help. Environmental rules are complex, and because of funding cuts we have few staff to explain them. After all, rules are generally one-size-fits-all. With an understanding of the real experience of being regulated, the assistance programs can become important sources of information about how regulations can be tailored, revised, and improved. The two hands can communicate about such things if there is a head as well. I did not always find that my reports on regulatory problems were well received by enforcement personnel, some of whom were excessively defensive and resistant to criticism. This is understandable because of the harsh and unrelenting and often unjustified criticism that has been directed at them. But a central coordinating function that upholds both activities can bring us the benefits of both.

REVERSING THE CURSE

There is a strange dynamic in environmental policy around the concept of the "unfunded mandate," which took shape in the 1990s around the Safe Drinking Water Act. Clinton's proposed tightening of standards for arsenic, based on new understanding of its hazards, prompted fears that small towns with small budgets for water departments would be burdened by the increased need to test. The correct response to this problem is obviously to do what Congress did in the days of the original environmental laws: to provide the necessary funding so that all towns—including small ones with small budgets—do the testing necessary, and set the standard according to what the science tells us. But the anti-regulatory conservatives who now had power complained that the federal government was imposing costs on the small towns and making victims out of them. They called the new standards an "unfunded mandate." The idea that the federal environmental effort was unfair settled like a curse on the mission of giving people clean water to drink. George W. Bush responded to this when, in one of his first acts, he downgraded Clinton's proposed standards for arsenic in water.[45]

The concept of the unfunded mandate is useful, though applied to the Safe Drinking Water Act it is simply a reminder that the mandate must be funded, not a reason to allow people to experience arsenic or other poisoning. If government requires something of people, should it not provide assistance in complying with whatever it requires? This seems like a basic moral principle.

45. He later reversed this decision, https://www.iatp.org/news/bush-administration-adopts-clintons-arsenic-rule-arsenic-standard-in-drinking-water-will-be-10-.

Instead of calling their own failure to fund the Safe Drinking Water Act an unfunded mandate, Gingrich et al. should have supported funding assistance to communities. We can reverse the curse by understanding this simple point, and we continue to live with it if we accept the idea that the unfairness comes from the recognition that levels of arsenic were too high.

Neither the affected community nor the regulated community should have costs imposed upon them. The affected community deserves protection, and the regulated community should have help. The mandate is morally and logically there, and it should be funded. As opposed to many investments we make as a society, this is an approach that has widespread and long-lasting benefits. We cannot escape the mandate to reduce the poisoning of our bodies, soil, water, air. This is not something we should think we have an option about. It is a mandate from Mother Nature, a mandate of Good Sense.

The devil's bargain of the Republican strategy of stoking anger at environmental governance has prevented its reform. At a meeting in the late 1990s, I asked why we shouldn't join in an effort to propose a fix to hazardous waste law, because *everyone in the room—representing business and environmental interests—was in agreement that the fix needed to be made and could be made*. It would help businesses as well as furthering the environmental mission. (It involved making recovery of wastes easier to do—this is still a very big issue in the hazardous waste world.) I was told we couldn't think of making legislative changes. If the environmental statute were put in issue the Republicans would try to repeal the whole thing. That was when Newt Gingrich was running the House. So you couldn't fix the problem everyone agreed could be fixed. The conservative antienvironmentalists were making regulatory *improvement* impossible. This is another story that should be better known. The businesses who cheered the conservatives cutting environmental programs were hurting themselves.

We need and can have systems that sustain life, our hopes and dreams. It is through a sensible system of self-governance that we can do this and have the maximum freedom we desire. Only an engaged, relational practice can give us the optimal balance and energy we need to find.

THE UNNECESSARY ACCEPTANCE OF RISKS

As I write this, workers are installing new windows in my house. I see the fiberglass fibers lit up by the sun as man pulls bits of insulation out. I ask, "Where's your mask?" He was just wearing one in my house to protect from COVID. He laughs and says this stuff is the good stuff, and he tears off a little piece and makes like he will rub it on his bare arm (but doesn't). "The yellow stuff, that's the bad stuff," he says, with the pink fiberglass cotton candy in his glove. I say I can see the fibers lit up by the sun, floating around him, and he should put the mask on. The fiberglass fibers are not good to breathe. "What's the yellow stuff you're talking about?" He says, "That made me itchy!" and says he can look it up. He takes off his glove and takes out his phone and after a bit he says, "Asbestos." I say, "Asbestos! Did you wear a mask then?" He says yes. I say, "Good you didn't breathe it!" The man speaks good English, I would say he has been here more than ten years. They gave him asbestos to remove and didn't take care to make sure he didn't get it all over his arms. Did he bring it home on his clothes?

We can go to the store and buy chemicals and put them all over our lawn and kill birds and insects and make it necessary to go out and buy more chemicals so that our lawns can grow after everything in them has been killed, and the government is discouraged from urging you to be aware about how you are risking the lives of your children, pets, and yourself, as well. The discouragement occurs in so many ways we don't see, with indications of political opposition, with shifts in support and funding, with public messaging, with studies to support a point of view, with claims made and asserted.

Today, Household and Commercial Product Association, the group that suggested they might sue us (the state) for recommending people stop buying products that became hazardous waste, continues its work. Its 2019 annual report states,

> HCPA successfully defeated proposals aimed at restricting the sale of neonicotinoids in Oregon, Minnesota, Illinois, New York, Virginia, New Jersey, Hawaii, Texas, and North Carolina. Legislation is still pending in Massachusetts.
>
> HCPA and the American Cleaning Institute (ACI) initiated litigation in October 2018 to challenge New York's Household Cleansing Product Information Disclosure Program, issued by the New York State Department of Environmental Conservation (NYSDEC).
>
> 1,4-dioxane remains a hot topic in New York, and this year, the state legislature passed a bill banning the sale of household cleansing products containing more than two ppm of the non-intentionally added ingredient. If signed by Governor Cuomo, this ban would take effect on December 31, 2022, with the threshold dropping to one ppm on December 31, 2023. HCPA staff has met with the Governor's office multiple times to discuss this issue and offered a chapter amendment proposing an industry-supported ban on household cleansing products containing over 10 ppm of 1,4-dioxane.[46]

---

46. https://www.thehcpa.org/wp-content/uploads/2019/12/2019-GRPP-Annual-Report-final.pdf.

What's confusing to an observer is that these actions, against state programs, do not appear until halfway through the report, which is preceded by seemingly cooperative initiatives celebrating such things as green chemistry Safer Choice products and a national ingredient standard. It is fine for corporations to argue the case for a national standard, even to say that we can meet 10 ppm, but not 2, for 1,4-dioxane. But to work to prevent states from doing what they see as necessary to do—citizens need to realize that the corporations are asserting a right greater than that of the states—is arguably unethical participation in democracy. The idea of the state's police power, used to protect public safety, is original and continuing in our traditions. Brandeis's idea of the states as laboratories of democracy is a vision of a society that can evolve sensible governance. This corporate vision of having one party to deal with—the federal government—and to render the states impotent concerning their products and their impacts by preempting their ability to regulate—is stifling.

Government is off-balance. Chemical companies can hurt the chances of elected officials. Oil companies can make them. Tech companies have "sway." Hedge funds have made their way. We need the separation of business and state, the same as we have tried to effect concerning the relationship of religion and state.

WHAT WE CAN HAVE

The story that needs to be told is what we can have. We can have a world in which workers are not harmed because of cruel economic demands. We can have a world in which the government helps you stop buying those chemicals and poisoning your lawn and children.

We can have a world that

- incorporates pollution prevention into the permit process, so that companies seeking permission to pollute have to pursue prevention first;
- provides greatly expanded assistance, education, and incentives for cleaner alternatives;
- fosters greater research and demonstration where toxic chemical use stubbornly persists;
- helps companies anticipate avoid "regrettable substitutions," in which alternatives turn out to bring new problems;
- helps companies understand the health and environmental impacts of material choices;
- extends what we learn from companies to consumers, so that they can vote for cleaner products with their spending;
- understands resource availability and the consequences of extraction;

- reviews all existing and proposed regulations for their preventive potential;
- expedites regulatory changes that better foster prevention;
- better rewards cleaner alternatives that work just as well or better than toxics, enhancing market forces where they don't do the job alone;
- better anticipates post-use impacts and doesn't just look at risks during use;
- better understands life-cycle impacts, including from raw material development through transportation and final disposition, including impacts on natural cycles and other life-forms.
- discards the practice of discounting future benefits for the sake of current monetary value;
- makes it easier to obtain compensation in court for harm caused by those who recklessly place toxics in commerce;
- does not wait for harm to show up before acting, but continuously improves the ability to anticipate and avoid it;
- creates powerful differentials between businesses that act responsibly and businesses that don't;
- understands and addresses international trade aspects of harmful commerce, bringing about transnational accountability; and
- builds the foundation and structure of the system of reliable independent science on which all of the above depends.

This is the story that needs to be told: of what we need not accept and what we can have. We have to ask for these things and elect only adults who will listen and work to produce them for us.

In my class Developing Sustainable Environmental Responsibility, I often ask my students what they do to be more environmentally responsible, personally. After we all go through that on our own and share what we do, we put that together with the list of problems we made at the beginning of the class.[47] If everybody did what we did, would it be enough? Not even if they all do all the things we do. It still doesn't come close. The exercise causes us to lift our eyes to the things we could do together, to the vision of what our united voices can do, our engaged planning. We can together focus on

47. These run the expected gamut from using less plastic and wasting less food to biking or walking and eating less meat, and rarely include knowing all the ingredients in things we buy so we can patronize the companies who make things without toxics, and their environmental practices, and how they treat their workers, and how they treat the land, so we can vote with our dollars. This is how capitalism could work, but such a system is not very well in place, though some good starts are being made (and sometimes co-opted). Thinking about things we need, such as the information tools that could empower the vast majority of those who would rather buy from the responsible, who recognize that attention to responsibility goes with attention to quality, can be one of the benefits of performing the exercise that shows how we come up short. The learned lesson, delivered by a cynic, can be helplessness. But framed as assistance to gaining the accurate perception of things so problems can be effectively addressed, and followed with creative redesign visualization, it can be empowerment.

the future of our children, ourselves, all living things. It is not a job just for some subgroup. We need the meeting of the minds that forms a new social contract, that this world is worth saving, and each of us has a part in saving it. We can give up the fights that spring up, because of history and being crowded, between ourselves. We can take control of this Earth, and agree to share it, and make sure it grows in good health. It is a matter of choice, not fate, if we continue tearing at each other and tearing it apart. We can agree to be members of a healthy body politic. The choice is in our minds.

RECOMMENDED ACTIVITY

Research whether your state, country, county, or municipality has had a pollution prevention program at any time. Research what assistance government provides for energy or water conservation. Find out what exists to help people recycle. Ask whether these programs can be reinstated or expanded.

Chapter Four

The Story That Poisoned Our Well

Thinking broadly about environmental issues includes considering the workplace. One aspect of how the workplace affects people is ergonomics, which means fitting the job to the person. Many of our problems stem from pursuing practices that don't quite fit the living, breathing reality of our bodies and the living world. The story of the ergonomics rule illustrates much of what we have been living through for a few decades now: the backlash against the public interest movement of the 1960s that established environmental, consumer protection, and public health laws began immediately and built over time. It was born and nurtured in offices of corporations, conservative academics, think-tanks, law and public relations firms, but has festered in angry populism. A straight account of our recent history recognizes a general hostility to laws that protect people has been purposely generated, in service to what can only be termed "special interests" such as coal, chemicals, agribusiness, oil, and plastics.[1]

We see the culmination of many years' work in the violent deregulation of the Trump administration, but we should see this trend from its roots, from

1. One feature of how advanced the practice of smearing public interest law has become is the success with which public interest efforts have been branded as "special interests" by those working in the interest of certain owners, managers, and organizations. This is a subversion of reason that can be undone by recognition.

the Powell Memo[2] to the creation of the Koch Networks,[3] and know how it fits into a powerful strand in the historical record, resonating with themes from the Southern Rebellion,[4] the Robber Barons,[5] and the resistance to Franklin Delano Roosevelt's[6] expansion of government.

THE ATTACK ON OUR AGENCIES

By 1996, Congress had enough representatives working for the part of industry that resents government interference to pass the Congressional Review Act, which made it easier for Congress to overturn agency rulings. For polluting companies, agency regulation can be seen as a cost: the EPA can feel like an enemy of profits. Call this the narrow view. A larger view sees that it is all of us who are threatened, and effective environmental law is something we all need. The larger view sees that the costs of not protecting are very great, but when you narrow your view to your own interests, you just don't see these costs.[7] The narrow view sees no lack of patriotism in attacking an instrument of our democratic system, it is just business to litigate, lobby, and argue whatever will turn the tide running against your revenue plan. "All's fair in love, war, business and politics." It even seems like patriotism when

2. From attorney Lewis Powell, soon to be nominated to the Supreme Court by Richard Nixon, to the education director of the US Chamber of Commerce, advocating corporate political action to counter "attacks" on the free enterprise system. Nancy MacLean's 2017 *Democracy in Chains* describes it as "widely cited as the beginning of the corporate mobilization to transform American law and politics." (125). See the memo at https://reclaimdemocracy.org/powell_memo_lewis/.

3. Jane Mayer's 2016 *Dark Money* describes how two brothers, both ranked among the top-ten wealthiest men in the world from fossil-fuel ventures, quietly "subsidized networks of seemingly unconnected think tanks and academic programs and spawned advocacy groups to make their arguments in the national political debate. They hired lobbyists to push their interests in Congress and operatives to create synthetic grassroots groups to give their movement political momentum on the ground. In addition, they financed legal groups and judicial junkets to press their cases in the courts." Mayer notes that the Kochs were not the only ones using their wealth to carry out political action, "often with little public disclosure," but that they were joined by Richard Mellon Scaife, Harry and Lynde Bradley, John M. Olin, the Coors, and the DeVos family: "together they formed a new generation of philanthropist, bent on using billions of dollars from their private foundations to alter the direction of American politics" (3–4).

4. For example, the position of John C. Calhoun that the Southern states could "nullify" federal law they found obnoxious. See Nancy Maclean's 2017 *Democracy in Chains: The Deep History of The Radical Rights' Stealth Plan for America.*

5. For an exhaustive treatment of how laissez-faire concepts benefiting concentrated economic power became embedded in American law, see William Wiecek's 1998 *The Lost World of Classical Legal Thought.*

6. See, for example, Kim Phillips-Fein's 2010, *Invisible Hands: The Businessmen's Crusade Against the New Deal* or the story of Smedley Butler, https://berkeleydailyplanet.com/issue/2011-03-09/article/37447, https://www.npr.org/2012/02/12/145472726/when-the-bankers-plotted-to-overthrow-fdr.

7. Upton Sinclair said "It is difficult to get a man to understand something, when his salary depends on his not understanding it."

one swallows the fish story that the essence of Americanism is the freedom to make as much money as possible. The larger view sees in this behavior, this crowding out of common interests from the center of public focus, what economists call "bounded reality." If there are enough public-minded citizens in a system of free enterprise, it can get along, but if too many only care about their immediate context, we see the erosion of the social fabric that makes our system of laws work and a loss of collective capacity to manage our common property: this natural world.[8]

The Congressional Review Act (CRA) was a direct result of the false theory that regulations are generally not a good idea, that government is always inefficient, that business is best left alone. This idea is not held or promoted by all who have to comply with rules or think they are inefficient and intrusive. But it is held by enough for there to be a current epidemic of hatred and resentment of the agencies charged with protecting us all. Restoring effective environmental governance requires countering this basic threat to our well-being.

The CRA didn't really do much that Congress couldn't already do, and there are better tools for controlling agencies. It was a symbolic message, an act of division, a declaration of war from the right wing: we are going to get those terrible agencies! The trend of distrust of government did not peak when Ronald Reagan said, "government is the problem,"[9] it continued and worsened, especially after leaders like Newt Gingrich showed how you could gain power through the generation of resentment against the very tools of our system of mutual self-governance.[10]

The first use of the Congressional Review Act was to vacate a rule on ergonomics instituted by the Clinton administration, during the first year of the George W. Bush administration. The Occupational Safety and Health Administration (OSHA) had worked on the rule for ten years. They had

---

8. As an example, the day of this writing, May 2, 2021, the *New York Times* writes of Mitch McConnell's predicament: if he supports the infrastructure plan of President Biden, he can finally see the Brent Spence bridge between Kentucky and Ohio fixed. But as a Kentucky Republican explained, "On the other side, he's the Republican leader of a caucus that doesn't want to cooperate with Biden" ("Biden's Expansive Infrastructure Plan Hits Close to Home for McConnell"). One mustn't without good cause accuse anyone of bad faith, but the oppositional behavior of the Republican party has exercised a corrosive force in recent decades. Repairing the social fabric requires recognition of what is happening to it.

9. https://www.reaganfoundation.org/ronald-reagan/reagan-quotes-speeches/inaugural-address-2/.

10. A May 2, 2021, *New York Times* review of John Boehner's memoir, *On The House*, describes how Newt Gingrich "called on the party to abandon concerns about bipartisanship and civility in order to obtain power." For a graphic summary of Gingrich's impact, see A. J. B. Lane's "Dragon's Teeth," in the *Boston Globe*, January 17, 2021: "the seeds of insurrection were actually planted three decades ago, by this man . . . he slashed at Democrats with incendiary language, casting them as enemies and questioning their patriotism . . . Gingrich's speakership coincided with the birth of Fox News Channel, which provided a welcoming new home for his inflammatory message," https://www.bostonglobe.com/2021/01/14/opinion/dragons-teeth/?event=event12.

figured out that it would prevent 4.6 million injuries and would cost industry about $4.5 billion a year in compliance.[11] Industry claimed it would be far more than that[12] and would put some companies out of business. "This regulation is an unnecessary burden and it got the death it deserved," said the spokesperson for the National Federation of Independent Businesses. *Yet all the rule required was that companies inform their workers about repetitive strain injury.* They didn't have to take any other action unless someone was harmed.[13] How do we understand this attitude? Should the industry be allowed to save some money by saying nothing about the risk and doing nothing about the harm?

Analysts such as David Dreisen have pointed out that a static picture of costs before a regulation goes into effect can be far higher than after a rule has been in effect and businesses have adjusted.[14] Scholars have shown how the estimated costs of rules have often been extravagantly overestimated.[15]

According to the Bureau of Labor Statistics, musculoskeletal disorders (MSD)—which can be prevented by ergonomics (fitting the job to the person)—account for fully one third of injuries in the workplace.[16] OSHA estimated that "$1 of every $3 spent on workers' compensation stemmed from ergonomic issues and that the direct costs attributable to MSDs were $15 to $20 billion a year, with total annual costs upwards of $54 billion."[17] Actions such as issuing guidance and greater awareness seem to have reduced the incidence of injury but per-case treatment costs have risen.[18] The narrow view would say businesses were saved regulatory compliance costs, but the larger view would see that had the rule not been overturned by a Congress representing regulatory alarmists, businesses would likely have seen much greater reductions than could be achieved with voluntary means only. In addition, the cumulative savings from 2001 till now—of twenty years of lower rates of injury—would have been substantial.

---

11. If you assume that the injuries cost an average of $1,000 per worker, the benefits exceed the costs. But imagine if you had an injury that makes it a misery or impossible to move a part of your body. Would you value avoiding that at a thousand dollars?

12. "The Mysteries of the Congressional Review Act," unsigned, June 1, 2009, *Harvard Law Review* 122 (2009): 2162, https://harvardlawreview.org/wp-content/uploads/pdfs/vol_122_the_mysteries.pdf. The claim was that it would cost more than $100 billion.

13. https://www.spineuniverse.com/wellness/ergonomics/ergonomics-rules-oshas-final-ergonomics-standard.

14. *The Economic Dynamics of Environmental Law* (2003).

15. https://www.pewtrusts.org/en/research-and-analysis/fact-sheets/2015/05/government-regulation-costs-lower-benefits-greater-than-industry-estimates;http://frankackerman.com/publications/costbenefit/Unbearable_Lightness_Regulatory_Costs.pdf.

16. https://www.osha.gov/ergonomics.

17. https://www.ehstoday.com/standards/osha/article/21919092/osha-and-ergonomics-the-past-present-and-future.

18. NIOSH ergonomics study.

If you are working, how you interact with things is the immediate environment in which you function and the benefits of a healthier workforce spread out into the societies we live in, constituting a feature of our human-made environments. If you find you cannot continue working as you did before, because now your arm is killing you if you move it a certain way, then how you function in your environment—how you relate to it—must change. Attention to the *relationship* is the beginning of solving the ergonomics problem, and also the beginning of responsibility. Seeing how we relate to our immediate environment; how animals, plants, soil, water, air, and food relate to each other in an ecosystem; and how we relate to each other in society, we have a chance of finding the answers to the threats to our collective selves. In order to find and grasp these answers we have to maneuver between small and large contexts, and allow our perspectives to change. Some may feel uprooted by taking a larger view—or empathizing with another's—and have to be helped to step out of their limited perspective. Ergonomics is rooted in its human focus, but we must begin with ourselves. In asking the question, what are we doing to ourselves, we begin a process that can be applied to preventing other impacts we cause.

A larger view encompasses both the point of the view of the trade association lobbyist hoping to save members some money and the point of the view of the suffering worker, and of the facts, including that getting rid of the ergonomics rule did not get rid of the ergonomics problem. Workers and businesses were and are still harmed by it. What's more, OSHA can still enforce against a business that caused MSD because it has powers under something called the General Duty clause, which requires a safe workplace. Having no explicit rule means a business has to figure out on its own how to do what it needs to do. Having a rule would have helped businesses know how to be in compliance and thus be safe from any enforcement by OSHA. Deregulation of this rule has not helped the regulated entities,[19] much less the vulnerable worker.

When facing the wreckage that the recent administration has made of environmental regulation immediate and long-term tasks present themselves. Some discussion in the policy world as the turnover to the Democrats took place involved whether they can or should use the Congressional Review Act to undo the most recent damaging regulatory actions. One prominent example was the rule about how scientific evidence is to be used that former chemical industry lobbyist and Trump EPA administrator, Andrew Wheeler, promulgated just before leaving. This rule, in the name of "transparency," is actually

19. The General Duty rule is rarely used, although it is widely applicable. But because it is sometimes used, the regulated population would be better served by an ergonomics rule that clarifies exactly what's required.

a rule about *not* using scientific evidence, instituted so that new evidence of significant harm from particulate matter could be ignored. This was a particularly violent attack on science in the service of companies who would have to reduce emissions that the science would show need to be reduced.[20]

Another example, less covered in the news, is the rule dictating how cost-benefit analyses are to be done before a rule can take effect,[21] which again in the name of something that makes sense, understanding costs and benefits, is actually a means of stopping the promulgation of regulations by providing for court challenge of every step in the process. There is a name for when well-meaning policymakers don't make decisions but keep studying the problem: paralysis by analysis, but usually that refers to an unintentional effect. Cost-benefit analyses, invented to help us make smart decisions, have been used for years to slow regulatory development. This attempt was to fully weaponize it to freeze decision-making. This is just one of the lesser-known disasters that we avoided by not electing Donald Trump to a second term.

The far-right pro-polluter lobby has succeeded in causing great damage. People need to see the chains that have been thrown on the great workhorses that are our environmental agencies to prevent them from working on our behalf. We must free them to do our work for us. We must not be afraid of the corporate behemoths and sly agents that have wrought this situation. Forgive them if you must, but recognize that it is *our* acquiescence to the logic of the dollar that sacrifices necessity. Environmental and other public interest agencies are property of all the people and we the people need to assert this fact until the danger recedes. Asserting the rights of the whole people can eliminate the dominance of special interests who disparage the needs of the rest of us. Quite simply, some large corporations are like beasts who are useful if tamed, destructive if allowed to run wild. How can they be tamed? By public recognition of the need and understanding of the tactics they are using to prevent our taking control of the context of our lives.

The Congressional Review Act powers are temporary, and Congress must act fast if it is to use them.[22] But it must also consider whether it should forgo using the CRA powers to eliminate rules made by the previous administration and, instead, repeal the Congressional Review Act. The big picture

20. The Transparency Rule was vacated February 1, 2021, by a federal judge because it was substantive and not a "housekeeping" rule, and had to be promulgated with notice and comment under the Administrative Procedure Act. http://blogs.edf.org/climate411/files/2021/02/38-Order.pdf.

21. "Increasing Consistency and Transparency in Considering Benefits and Costs in the Clean Air Act Rulemaking Process." See Richard Parker and Amy Sinden, "Andrew Wheeler's Trojan Horse for Clean Air Act Regulation," *Yale Journal on Regulation*, January 6, 2021, https://www.yalejreg.com/nc/andrew-wheelers-trojan-horse-for-clean-air-act-regulation-by-richard-w-parker-amy-sinden/.

22. This was written before the Democrats used the CRA: https://thehill.com/opinion/white-house/556133-democrats-use-of-congressional-review-act-puts-filibuster-debate-in-new/. These efforts were to control agencies out of control because of a White House out of control.

suggests this. Using it would reify it, and some say we should just get rid of it if we can.[23] The CRA was a reflection of the idea that Congress needed more power over regulatory agencies, reflecting the view that these were great beasts running riot (call it confusion about who is the beast). Yes, the administrative state can become a fearsome beast. But it is also our bulwark against concentrated selfish power. As noted in an Administrative Law Note in the *Harvard Law Review*, "The problem of congressional control of the administrative state is not new. As early as the 1930s, members of Congress worried that wide delegations of administrative authority would leave the unelected bureaucracy politically unaccountable. Yet they also realized that Congress could not pass enough specific legislation to regulate the increasingly complex world."[24]

Agencies are created by Congress, to accomplish an important public purpose. Once the government of the people has done this, is it a good idea to make it so easy for a subsequent Congress to overturn what the agencies do, without going through the public engagement that the agency must go through? There is insufficient appreciation of the fact that the 1946 Administrative Procedure Act makes agencies provide the public with notice of any rules they wish to develop, before they create them, and the public has the right to comment on them.[25] The public gets to speak at hearings on the rules, and has more chance of interacting and perhaps influencing the shaping of the rule than many people understand.[26] Our regulatory process is basically open and democratic. But the conservatives talk about it as if it is tyranny. The CRA comes out of that twisted story, and the truth is that it puts in place a less democratic process, and a less thoughtful one.

The law was introduced by conservative Republicans to wipe out rules that their supporters didn't agree with, and the first time they used it they wiped out a rule that would have saved a lot of people from injury and businesses from avoidable costs. In mere days, they undid work crafted over many years. Ergonomics remains a problem. Data from the Bureau of Labor Statistics shows that "these types of injuries account for one-third of days-away-from-work cases. The BLS data further explains that employees suffering from

---

23. Thomas O. McGarity, Rena Steinzor, James Goodwin, and Katherine Tracy, "The Congressional Review Act: The Case for Repeal," Center for Progressive Reform, 2018, https://cpr-assets.s3.amazonaws.com/documents/CRA_Repeal_Case_050218.pdf.

24. "The Mysteries of the Congressional Review Act," 2164.

25. I state, as evidence of this ignorance, my experience teaching law since 1990 to thousands of adults of all ages.

26. Many agencies have advisory groups and conduct events to get public input. Many agency staff are eager to meet with concerned citizens to discuss policy—as are legislative staff. In my experience receiving comments from the public prior to writing final versions of a regulation, I was never told to place more weight on one comment than another, nor did I witness this. Instead, we discussed the merits of the points made.

ergonomics-related injuries required more time off the job than those with other types of workplace injuries and illnesses (a median of eleven days versus eight days)."[27]

Obama vetoed several attempts to use the CRA, but by the spring of 2018, the Republican Congress and Trump had used it to void fifteen rules:

> safeguards that protected safe drinking water; kept weapons out of the hands of people suffering from mental illness; secured the privacy of Internet users' browsing data; allowed consumers a realistic way to fight back when cheated by banks, credit card companies and other financial institutions; promoted greater access to family planning and other health care services for women in low-income and under-served communities; and more.[28]

The Center for Progressive Reform found that

> The CRA's design renders it a poor tool for conducting meaningful congressional oversight. In particular, the shortcuts in legislative procedures it creates preclude, rather than advance, the kind of meticulous investigation and thoughtful deliberation that is at the heart of effective oversight. For the 15 rules that were fully repealed using the CRA's procedures in 2017, the process took an average of only 48 days to complete. This lightning-quick pace left little time for adequate investigation and deliberation. This quick process for repealing rules also stands in stark contrast to the long period of time agencies spent developing them. On average, the 15 rules that were eliminated through the CRA had been in the works for approximately three years each.[29]

We can argue whether Congress should have more control over agencies. But we also need stability, and change when needed done thoughtfully, after engaged deliberation that seeks consensus. That's what regulatory process, done well, achieves. The narrative that justified the CRA was false, that it was something for the people. It was part of the Contract with America,[30] which was essentially saying to the American people that their own government had been taken from them and the conservatives were going to rescue it back for them. But they were really working in the interests of companies that wanted to shrug off regulatory controls, which was a backlash to the popular will that took the form of federal law that insists that these very companies must not excessively pollute our world. The contract had a small grasp on a truth—that the regulatory governance system has grown powerful and complex and oversight is necessary. But it polluted our discourse with the untrue story that the

27. https://www.ehstoday.com/standards/osha/article/21919092/osha-and-ergonomics-the-past-present-and-future.
28. "The Case for Repeal," 2.
29. Ibid., 4.
30. https://www.britannica.com/event/Contract-with-America

agencies are out of control and doing damage to America. It was damaging to our system of mutual self-governance.

The contract with America was written with the Heritage Foundation.[31] For just a sample of information indicating the foundation's resourcing see Sourcewatch.[32] In 2016, Senator Sheldon Whitehouse (D-RI) and eighteen other US Senators delivered a series of speeches denouncing climate change denial. The Heritage Foundation was one of thirty-two organizations doing this work for the fossil-fuel interests. Whitehouse intended to

> shine a little light on the web of climate denial and spotlight the bad actors in the web, who are polluting our American discourse with phony climate denial. This web of denial, formed over decades, has been built and provisioned by the deep-pocketed Koch brothers, by ExxonMobil, by Peabody Coal, and by other fossil fuel interests. It is a grim shadow over our democracy in that it includes an electioneering effort that spends hundreds of millions of dollars in a single election cycle and threatens any Republican who steps up to address the global threat of climate change. . . . [I]t is long past time we shed some light on the perpetrators of this web of denial and expose their filthy grip on our political process. It is a disgrace, and our grandchildren will look back at this as a dirty time in America's political history because of their work.[33]

RESPONSE TO THE ATTACK

The perspective of the company that supplies toxics is not the same as the interest of the company that uses them. The latter may be annoyed by regulations that pertain to the use of those toxics, but if that company can find a substitute, it is no longer burdened by any regulation, and is free to create a safer environment for its workers, neighbors, customers. Similarly, the workers and the union that represents them may fight regulation that threatens their jobs, but if we have transitional policies in place, this problem can be lessened. Companies using toxic products would be happy if their raw material supplies were not toxic: it is products they want to make, and workers would be happy if their jobs were not toxic. Both may perceive anti-toxics law as an economic threat, but they might be happier if they can implement a transition. *It may only be the seller of the toxic material who is threatened* by laws meant to prevent the harm their sales are causing. Though sellers of

31. https://www.heritage.org/political-process/report/the-contract-america-implementing-new-ideas-the-us.

32. https://www.sourcewatch.org/index.php/Heritage_Foundation.

33. https://www.whitehouse.senate.gov/news/release/senators-call-out-web-of-denial-blocking-action-on-climate-change.

products extracted from the earth have a key role in our economy, in seeing the differentiated interests we can imagine a sophisticated response.

The small businessperson who has a hard time keeping up with all the regulations is not the same as the large oil or chemical companies. The Clinton-Gore administration deemed it morally sensible to have policies that help those who do not have their own legal department, in complying with rules. They began a process of making the rules more comprehensible, and funded assistance programs in the states.[34] They advanced the process of looking at each industry sector separately and tailoring rules to each context, not regarding industry as a monolith, and seeking those who would cooperate.

They reformed enforcement, strengthening it while also making it more humane, bringing into the process the means for recognizing good faith, making amends (the Supplemental Environmental Project),[35] self-reporting,[36] and Project XL,[37] intended to provide ways around rules that prevented desirable actions.

This didn't succeed in sufficiently reconciling the public to the existence of environmental enforcement, which remains the target of a simplistic antagonism toward government itself.[38] The CRA was passed as result of an unsupported belief that regulations harm us, but it is important to note how many legislators professing that position benefited from the contributions of powerful polluters.[39] Much of the voting public has now become convinced

34. Al Gore was in charge of implementing Bill Clinton's National Performance Review (later renamed the National Partnership for Reinventing Government), intended to reduce red tape and improve agency performance. An observer later wrote: "What many recall as the twentieth century's most visible effort to remake the federal government endures in the twenty-first in bits and pieces. That's because two succeeding administrations picked up only the components that fit their style, the march of technology rendered some revolutionary ideas quaint, and the public's respect for government, polls show, has trended nowhere but down," Charles Clark, senior correspondent, "Reinventing Government—Two Decades Later," *Government Executive*, April 26, 2013, https://www.govexec.com/management/2013/04/what-reinvention-wrought/62836/.

35. https://www.epa.gov/enforcement/supplemental-environmental-projects-seps

36. https://www.epa.gov/compliance/epas-audit-policy

37. See archives of this discontinued effort at: https://archive.epa.gov/projectxl/web/html/index.html

38. "Two Decades Later" includes the comment by Paul C. Light, a professor of public service at New York University, that "When George W. Bush entered the Oval Office in 2001, there was a 'noticeable chilling' toward reinventing government . . . a return to the Reagan-era notion that 'the problem with government is government.' The Bush team didn't like bureaucracy and favored outsourcing."

39. For an extended exploration of this, see Jane Mayer, *Dark Money: The Hidden History of the Billionaires Behind the Rise of the Radical Right* (2016); or shorter pieces: Suzanne Goldemberg, "Secret Funding Helped Build Vast Network of Climate Denial Thinktanks," *The Guardian*, February 14, 2013, https://www.theguardian.com/environment/2013/feb/14/funding-climate-change-denial-thinktanks-network; David Rosen, "Don't Be Duped by Corporate Spin: Regulation Protects Us All," November 19, 2018, https://truthout.org/articles/dont-be-duped-by-corporate-spin-regulation-protects-us-all/; and Tony Carrk, "The Koch Brothers: What You Need to Know," Center for American Progress, April 4, 2011, https://www.americanprogressaction.org/issues/economy/reports/2011/04/04/9399/the-koch-brothers/.

that regulations are generally harmful.[40] Although some advances were made during the George H. W. Bush administration (such as the strengthening of the Clean Air Act as well as the Pollution Prevention Act), and many great environmentalists have been Republicans, the extreme deregulation we have seen during the Trump era was the culmination of a trend building in that party over a long period.

Because fostering anger toward environmental and other public interest agencies worked for some to get elected, throughout the Obama administration citizens of this country have had to endure watching extraordinarily hostile questioning of our public servants, as the noxious myths about the usurpation of powers by agencies have grown to ugly extremes.[41] QAnon conspiracy stories are not that far from the cynical ploy of the Congressional Review Act. The loss of faith in our democracy was nurtured and administered—it didn't just happen as a result of bad government. Disregarding the great fact of modern times—that we are billions on this planet and need agencies to help us navigate the many complexities—instead of supporting the growth and improvement of these agencies too many Republicans have worked to retard the evolution of smarter systems. For progress to resume, we have to see this perspective for what it is and repudiate it.[42] For fairness it must be mentioned that there have been many disappointing Democratic positions but on the whole their history is one of supporting the democratic and ecological systems on which our prospects for health and happiness depend. We don't help things by avoiding stating political facts. How do we rescue and bring back the grand old party unless we recognize how far it has fallen from the Tree of Liberty?[43]

40. "'[f]or the 12th year in a row, more Americans say there is too much government regulation of business and industry than say there is either too little or the right amount.'. . . . Sixty-eight percent of Republicans compared to 20 percent of Democrats said there was too much regulation," Karlyn Bowman, "Where Is The Public On Government Regulation?," *Forbes*, October 23, 2017.

41. As just one illustration, see Rep. Paul Gosar angrily berating EPA administrator Gina McCarthy, https://www.youtube.com/watch?v=d1G-u7Qj-3Y. While the EPA deserved much criticism for its handling of the Flint Water Crisis, simply piling obloquy upon EPA does not help but serves another purpose.

42. To a Republican this may sound partisan, but history bears out the alignment of that party in recent years with antienvironmental and anti-governmental positions. I would be among the many who would celebrate the return of a responsible Republican party and applauds the bravery of those who stand against these trends.

43. It is possible that these thoughts can be lifted out of context and used to accuse me of being partisan. It is, however, necessary to face what's happening. I have worked with great Republican environmentalists. I appreciate many traditional Republican ideas. That is not an excuse for overlooking the civic destruction the party is causing, as of this writing, and the failure to look at it that allows it to continue.

ON NOT USING THE MEAT-AXE

The CRA is a crude axe-swing, not a careful fix. Congress created the agencies to carry out duties important to the American public. Criticism of regulation has mounted to the point that regulations are portrayed without having any benefits at all, as in the 2017 report of the Commerce Department on the need to reduce the burdens of regulation.[44] That view has led to the loss of important environmental and public health and public interest law. The Consumer Product Safety Commission has a near ghost-like status, never truly powerful enough to protect the consumer; OSHA has been hobbled, and EPA was turned into a villainous enemy of people. Not long after the Consumer Financial Protection Bureau was created it was virtually neutered. When we see that these are the things of the people and they have been taken away, we can begin the work of restoring them. If our history is understood, perhaps resumption of progress can be quicker than we think.

The history of environmental regulatory policy can be very dry. But in fact, it is as volatile as a crime story or novel of passion. We should see the drama of efforts like the CRA. We all benefit from comity and the stability when successive Congresses have some respect for what previous Congresses have created, and for the work of the agencies under them, and don't use a meat axe to fine tune things.

The *Harvard Law Review* note on the CRA describes how the overturning of the ergonomics rule looked to those who had been working with so many others for so long to develop it:

> Just before Congress disapproved the ergonomics rule, Peg Seminario, the AFL-CIO's director for health and safety, spoke to the New York Times and demonstrated the extent to which traditional notions of administrative effectiveness and democratic accountability have become inverted: "What's happening is stunning. . . . This rule is 10 years in the making, with 10 weeks of public hearings on it, and now they want to wipe it out with not even one hearing and less than 10 hours of debate. That's about as undemocratic a process as you can get." Seminario describes as "undemocratic" the majority vote in two houses of Congress and presentment to the President of the disapproval resolution. She contrasts the slow and deliberate regulatory proceeding with the fast and—in her view—ill-considered legislative process. Implicitly, she views the regulatory process as more democratically legitimate than the constitutionally prescribed legislative process.[45]

---

44. *Streamlining Permitting and Reducing Regulatory Burdens for Domestic Manufacturing*, https://www.commerce.gov/sites/default/files/streamlining_permitting_and_reducing_regulatory_burdens_for_domestic_manufacturing.pdf.

45. "The Mysteries of the Congressional Review Act," 2174.

The story was supposed to be that the agencies are out of control—unrepresentative—but the agency's process was actually far more representative of the people's will. Where democracy occurs is critical. A study by the Center for Progressive Reform concluded that the CRA is:

> designed to short-circuit Congress's deliberative process and allow narrow partisan majorities to attack broadly popular public safeguards on behalf of politically powerful interests. . . . By unwinding the significant public health, safety, environmental, or financial security protections these safeguards would have otherwise delivered, each CRA resolution that is adopted boils down to a direct assault on the public interest.[46]

THE BATTLE OF STORIES

A similar conservative story is that the judicial concept of "deference" allows too much power to flow to the agencies. Courts are pressed to overrule agencies imposing rules on businesses, and that they should not defer to the agencies' judgment. It is a complicated argument for conservatives, because when agencies do things they like, they want courts to defer.

The idea of deference is that the agency was created to do a job and a court will not just step in and undo its actions unless there are good reasons to do so, basically that the agency has not obeyed its mandate from Congress, has been inconsistent in application of its own rules, or was outright "arbitrary and capricious." The public needs to understand that Congress can't do the job agencies were created to do, nor do we want courts stopping them from working when they are doing their job. Courts have a role, like Congress, in ensuring they work properly, and that's how they should be used.

The general concept that society has a right to effective regulatory agencies should not be in question, and it has been now, for far too long. Congress created agencies to serve us. They hire experts to develop the capacity to serve us. We paid with our taxes for the agency to do its work.

Sensible public interest regulation is of benefit to us. There is an extensive literature concerning this campaign, the promotion of the idea that government is the problem in our lives; that it restricts our freedom, and that

---

46. *The Case for Repeal*, 1.

environmental rules are bad for us.[47] It is a major cause of the movement to hobble or eliminate governance that we need to protect ourselves. We are the victims of a false call for freedom.[48] But our freedom, to have a government that addresses our problems, can be asserted as well.

Antienvironmentalists argue that regulations cost money far exceeding benefits, while the analyses actually tell the opposite story. This is often a game of cherry-picking. In the 2015 case *Michigan v. EPA*, Justice Scalia found that regulating power plants to control mercury emissions (extremely toxic to nearby residents) would produce $4 to 6 million in annual benefits, while costing the industry $9.6 billion, and concluded this was not appropriate. But Justice Kagan in dissent pointed out that EPA estimated the benefits as $80 billion per year, and that the regulation would save as many as eleven thousand lives per year. What was the difference? Scalia refused to accept that the regulations which would control mercury would also control other pollutants.[49] In order to come up with a small estimate of benefits he ignored reality. In addition, that rigid and selective application of cost-benefit analysis ignored that those living near power plants are in hot-spots of mercury poisoning, and that preventing this is of incalculable benefit to them, and should not be offset by savings to the power company, which has no right to poison them.

Simple logic should tell you that investments intelligently made in our health, and to preserve nature, are of lasting benefit to us all.[50] The story of the ergonomics rule is illustrative. A cost-benefit analysis was used to make the case that it was not worth doing. But this does not capture the story of people. Why should we not spend some money to educate workers about how to avoid repetitive strain injury? They have the *right* to be free of it. The idea that a company would be put out of business by informing their workers about the danger of repetitive strain disorder is an absurdity. A business should be seen as including its workers, and they cannot do their work if they are injured. The reasonable approach would be to regulate efficiently, not to

---

47. For example: Naomi Oreskes and Erik Conway's 2010 *Merchants of Doubt: How a Handful of Scientists Obscured the Truth on Issues from Tobacco Smoke to Global Warming*; Norman Ornstein and Thomas Man's 2012 *It's Even Worse Than It Looks: How the American Constitutional System Collided With the New Politics of Extremism*; the Union of Concerned Scientists' Climate Change Deception Dossier, https://ucsusa.org/resources/climate-deception-dossiers; David Rosner and Gerald Markowitz's 2002 *Deceit and Denial, the Deadly Politics of Pollution*; and read about the careers of Elizabeth Whelan and Steve Milloy countering the public's view of environmental and public health science.

48. Thomas O. McGarity, *Freedom to Harm: The Lasting Legacy of the Laissez-Faire Revival* (2013).

49. *Michigan v. EPA*, 135 S. Ct. 2699 (2015).

50. The value of country investments in preserving nature are at the heart of the movement for nations to include accounting for the value of preserving their natural resources in their economic planning. A well-designed regulation provides opportunity and spurs innovation and improvements in product quality as well as public benefits.

decide to do nothing: not to question whether we need to, but to understand how best to do it. The reasonable approach would be to recognize that we should have governance that gives us businesses that don't hurt workers, not businesses that treat them as replaceable cogs in a machine.

Our concerns, after protecting the innocent, should be whether businesses can be helped to transition to be safer places. We will need the coercion of authority in some circumstances, but to get progress, we should regard it as a last resort, and make primary the use of education (as the proposed ergonomics rule did), and many forms of assistance. We are in this environmentally unsustainable society together, and it is together that we will climb out of the traps we have created for ourselves.

It may be a lot to ask of people to follow the arcane news about agency activities, to read books so that they understand these issues, to listen to thoughtful conversation about this area of governance, to concern and involve themselves. But it isn't as hard as that just not to so quickly believe the lies that are flying around. The efforts of the public relations firms hired to convince people that government is bad, and the politicians who tell you they are going to office to fight for your freedom from your government, can be seen for what they are. People can learn to recognize that an argument against government may indicate a very special interest that craves greater profits. They can learn how to check things out so that they are not manipulated. If we can help people to stop adopting pernicious myths about our own democracy then the ones who have worked to capture it for their own private, not shared interests will have less power and we can be free to do the work that needs to be done.

As UCLA School of Law's Blake Emerson writes in *The Public's Law*, an examination of the role of administrative agencies, which can either supplant the public's role in decision-making or foster it, involves a "profound ideological shift":

> We must begin to think differently about the structure of the state—to understand administrative agencies not as instruments of presidential or legislative will but rather as agents of the people, directly. Public acknowledgment of the democratic authority of administrative agencies themselves would serve to ballast against the growth of presidential power . . . the procedures Congress has set forth in the Administrative Procedure Act provide a framework for an administrative state that draws its legitimacy from direct public participation rather than merely from the majoritarian mandate of the president or from the laws enacted by Congress. . . . A full renewal of the Progressive state would require deepening governmental intervention in society, coupled with

participatory procedures calculated to redress the structural injustices that often characterize social and private life.[51]

The concept of participatory democracy was articulated in the Port Huron Statement of the Students for a Democratic Society (SDS) in 1962, which, in the words of principal author Tom Hayden, was intended to empower "the individual as autonomous but interdependent with other individuals, and the community as a civic society."[52] In a participatory democracy, the statement read, "political life would be based in several root principles:"

> that decision-making of basic social consequence be carried on by public groupings;
> that politics be seen positively, as the art of collectively creating an acceptable pattern of social relations;
> that politics has the function of bringing people out of isolation and into community, thus being a necessary, though not sufficient, means of finding meaning in personal life;
> that the political order should serve to clarify problems in a way instrumental to their solution;
> it should provide outlets for the expression of personal grievance and aspiration; opposing views should be organized so as to illuminate choices and facilitate the attainment of goals;
> channels should be commonly available to relate people[53] to knowledge and to power so that private problems—from bad recreation facilities to personal alienation—are formulated as general issues.[54]

This vision of a democracy was labeled as a product of the "New Left" and the SDS as a radical organization opposing patriotic Americanism, because of its protest of the Vietnam War and racism. But it is consistent with Blake Emerson's vision of an administrative state responsive to the people, a concept he traces to American philosopher John Dewey (and before him to the German philosopher G. W. F. Hegel), and the respectable Progressive tradition, which was at one point headed by former president Theodore Roosevelt. "The state," wrote Emerson, describing the essential American version of Hegel's idea, would "attempt both to furnish the conditions for rational and inclusive discourse and to empower this discourse by including affected persons and groups in the policymaking process." Through such interaction with affected populations, agencies would "dismantle social relationships

51. *The Public's Law* (2019), 184.
52. http://tomhayden.com/home/participatory-democracy-from-the-port-huron-statement-to-occ.html. See also https://www.newyorker.com/magazine/2021/03/22/the-making-of-the-new-left.
53. The original says "men," but I assume Hayden would approve of the change.
54. Comment on the statement is easy to find: the statement itself is on the web archive, https://web.archive.org/web/20090123114146/http://coursesa.matrix.msu.edu/~hst306/documents/huron.html.

characterized by servitude, domination, and exclusion, and support in their place equal, integrated, and reciprocal relationships between citizens."[55] The story of Right versus Left that has so dominated recent decades obscures the fact that the essential ideas of participatory democracy are outgrowths of the seminal document of American freedom, the Declaration of Independence, and trace back to many of this country's founders, from both North and South.

At this writing, the battle of stories is clear in the *1776* response of conservatives to the *1619* recognition of liberals. As the *1619* series by the *New York Times* met with approval for facing history as it is, and the Black Lives Matter movement likewise, the Trump-commissioned *1776* report countered with praise of America. Ironically, the praise centered on its elevation of the concept of equality. While critics noted that did not eliminate the fact of slavery, this form of storytelling is what should be noted. It is what you end up with if you keep going after you begin to slant things. The Bushes said they were balancing jobs and environment, that both were important, but Dick Cheney's Energy Task Force was nearly completely fossil-fuel and was kept secret. This is not balance. Trump went even further, making poor toxic coal the victim of an unjust war that had to be beat back, and issuing so many deregulatory orders that Harvard, Columbia, and other law schools created "Deregulation Trackers" to help the public understand them.[56] He talked about tearing down regulations as an achievement. The simplicity of the conservative idea that liberals are awful leads to the QAnon fantasies that they are actually devil-worshippers. The temptation of the simple story, with its rewards of acceptance and approval by the in-group, can be as powerful as the strongest drug and as narcotizing of thought.

The work of agents of the development of responsible action involves being story doctors, diagnosing the bad stories at the heart of unfortunate political movements, noting how reductive thinking cuts through disbelief and discomfort. And they must be true story deliverers, helping those addicted to simplistic views and blind to the need for balance and complexity, to see and hear what makes sense and above all, to crave it rather than the jiggered Kool-Aid they've been drinking.

The stories that can make sense to anyone are available. They are the medicine we can use to cure the ills that bad stories have inflicted. It is one thing to cite the numbers that show the benefits of regulation. In 2011, *EPA's second evaluation of the costs and benefits of the Clean Air Act showed*

55. *The Public's Law*, 185.
56. https://www.brookings.edu/interactives/tracking-deregulation-in-the-trump-era/; https://climate.law.columbia.edu/climate-deregulation-tracker; and https://eelp.law.harvard.edu/2018/07/tracking-the-trackers/.

benefits exceeding costs by more than thirty to one.[57] While that information has helped to fend off efforts to repeal the act, we still do not have enough of a public that can respond to the simple logic that it's better not to breathe pollution. We shouldn't have to have studies to prove that. We should be able to say, like Cher in *Moonstruck*, "Snap out of it!" and get people to wake up.

At conferences filled with technical presentations on cost-benefit and risk-benefit calculations, I have had the opportunity of presenting a thought experiment. To rooms filled with scientists and technical policy wonks I asked the following question:

> Imagine that you are the host of a party at which you are serving wine that you are very proud to serve—extremely fine wine that you are generously providing to everyone. You have already talked about how great it is and everyone is looking forward to it.
>
> Just before you serve it, you are informed that it has been discovered that the wine contains one part per billion sewage. Do you serve it?

The result of this thought experiment was always a great deal of consternation. People said that one part per billion sewage was nothing to be concerned about. Some would say whether they would serve it or not. But most squirm uncomfortably and will not take a position.

Then I would ask a follow-up question.

> Do you have the responsibility to inform people of what you have learned?

This also created consternation. People were unsure. Most said they thought they did have that responsibility. Only a minority would say that they would withhold the information.

Then I would ask the last question.

> If you were a guest, and you were informed of this, would you drink it?

This always garnered expressions of disgust. The overwhelming majority has declined to drink wine with sewage, even when it is only one part per billion.

Risk and cost-benefit calculations are interesting. They do inform us of important facts. But in the end, what is really important is something else. The thought experiment—easy to do—is about relationships. Do you have the right to decide for someone else? The numbers say it's not of concern, but when you change positions, you shudder with disgust. If we can feel the

---

57. https://www.epa.gov/clean-air-act-overview/benefits-and-costs-clean-air-act-1990-2020-second-prospective-study.

difference between choosing for ourselves and choosing for others, we see that a risk-benefit calculation is misleading. We see others have that same right to know that we want for ourselves.

The decision-making process should be about respecting people first and foremost. The right policies on ergonomics and all the other issues we have to deal with are arrived at by considering relationships and how people should be treated, and how we should behave in this fragile world. This is a story we have to tell, to counter stories that have poisoned our well.

EXERCISE

Write down on a piece of paper three things that the government can do to counter misinformation and better inform the public. Think about how there is free speech and how the government should describe its role, how this can be done without establishing a government propaganda machine that could be misused. Then share your ideas with others. This can be an exercise in class or it can just be something for people to do, like a party game, but serious.

Chapter Five

The Big Picture

The one silver lining in the dark cloud of looming environmental catastrophe is that it affects us all and, thus, provides a clear and pressing reason to unite as one body of humanity. But universality can feel like a threat to sub-universal identification. We each define ourselves as particular kinds of human, distinct from others, but some define themselves in opposition to other kinds. The history of religion includes many figures who opened their arms to every person, but it also includes those who have made war on those who have different beliefs. In the modern era, religious wars may have lost some legitimacy, but they persist. The struggles for environmental responsibility waged in the halls of government begin in the places where our religious, philosophical, and cultural identifications come together to constitute a position in the world or a worldview. The very openness to worldviews, other than the one selected, must be established. Those who feel threatened by interaction can be helped to discover the benefits of respectful exchange.

Some fear the loss of a moral system through the expansion of our context to universality. When the Universal Declaration of Human Rights was being debated at the United Nations in 1947, the American Anthropological Association stated that "moral values are relative to cultures and there is no way of showing that the values of one culture are better than those of another."[1] This seems to assume that each viewpoint will be unchanged by encounter and we face a competition in choice, but we can instead create a space open to all manner of seekers.

1. Chris Gowans, "Moral Relativism," *Stanford Encyclopedia of Philosophy*, https://plato.stanford.edu/entries/moral-relativism/.

CONTEXTUAL EXPANSION SEEMS A THREAT TO SOME

Those who believe their faith is the sole truth may respond as Cardinal Joseph Ratzinger did (before he became Pope Benedict XVI), that "we are building a dictatorship of relativism . . . whose ultimate goal consists solely of one's own ego and desires."[2] Or, as the religious website "All About Philosophy" argues, "Once right and wrong are relegated to matters of opinion at worst, or are purely subjective at best, any talk of morality becomes incoherent."[3] It may be hard to see the silver lining of environmental awareness—which forces recognition of our basic nature as members of one species—an undeniable universality—in the destruction of limited visions of morality, but it is possible for those with faith that moral principles are inherent in the recognition of our shared reality.

Attempts to find objective or universal principles—as thinkers have tried to do since Aristotle, whether through narrative logic as Kant and others have tried, or through mathematics as Bertrand Russel tried, have thus far failed to unite the world. The modern philosopher David Copp proposes that the shortcomings of moral relativism are mitigated by the fact that all societies base their moral systems on fundamental needs, which are similar. Fordham University's Chris Gowans writes in the *Stanford Encyclopedia of Philosophy* that "Whether or not justified moral codes (and hence moral truths) would tend to be substantially similar, despite differences, as Copp argues, would *depend on both the claim that all societies have the same basic needs and the claim that these needs are much more important than other values* in determining which moral code it is rational for a society to select."[4] One may argue that the more environmental problems intrude on our basic needs, for breathing, food, clean water, or safety from floods, the more it will be more widely recognized how important indeed these needs are, relative to other values.

In a crowded world, where we can no longer form our own societies without extensive interaction and indeed interdependence on those who are different from us, the idea of moral relativism can be helpful if it increases our ability to tolerate differing beliefs. But if accepting some moral relativism means goes so far as to disturb long-held particular ideas about truth, how can we unite around a shared vision or task? I believe it is not a real problem, because the real problems put these ruminations into perspective. When a tidal wave is approaching, we stop arguing about how many angels can dance

---

2. Edward Sri of the Augustine Institute, in an article prompted by his book *Responding to Relativism with Logic and Love*, in the *National Catholic Register*, May 10, 2017, https://www.ncregister.com/commentaries/the-mask-of-moral-relativism?utm_source=bing&utm_medium=cpc&utm_campaign=ncrtraffic.

3. https://www.allaboutphilosophy.org/moral-relativism.htm.

4. 2019 edition, https://plato.stanford.edu/archives/sum2019/entries/moral-relativism/, emphases added.

on the head of a pin and we run to safety together. Not being drowned is an example of a universal value.

Having an ecosystem that sustains is a fundamental need. If we can help people to apply these perceptions, we have a chance at understanding environmental responsibility as an analogue of divine command, as if voiced by all the different gods and superseding all the logical constructs that have ever been devised to satisfy our need to have something to believe in. Until recently, only those with faith in science have seen the tidal wave coming.[5] Now it is hitting more and more people where they live and more are looking up and seeing the rising heat, the changes in the weather, seeing what people around them and they themselves are doing.

THE TEAM CONCEPT

Locating civic duties in the appropriately universal field—a necessity if we are to design our programs so that humanity's activities are compatible with the living Earth—leads to the problem of how you narrow down, how you get moving on things. Here the team concept, in which we combine our talents and energies, is essential, for the very large task requires many perspectives and capabilities. We can no longer afford to disappear into our various specializations, our provincial worlds. We must all now participate in a joint awareness of all the things that are necessary to know. This requires the division of the labor of understanding and a new level of trust in each other. The competitive mode must be subsumed. The relational mode, of each to each and each to all, must become primary. This is not dictatorship or oligarchy, for that will constrain capacity, understanding, and action. A healthy world democracy based on mutual care is our only chance of protecting life on Earth as we know and love it. Competition as the primary context denudes the world of value. It must be subsumed within cooperation—then it can be healthy competition, which is often known as "fun."

Though we are accustomed to forming teams to compete against other teams, as in the competition between corporate entities in the marketplace, or between teams of a particular sport, there is no reason that we cannot envision a worldwide team, in which every human is a member, which competes against the forces of death. In *The Science of Good and Evil*, Michael Shermer writes that while it seems that evolution would favor humans who are individually selfish, evolutionary selection actually favors groups that contain individuals with high levels of cooperation and altruism:

---

5. I am not counting those who interpret it as a fulfillment of a particular myth that confirms a separate version of reality asserted competitively.

> groups with more moral individuals than immoral individuals gain a collective advantage over other groups who are less successful at fostering a healthy balance between cooperation and competition, altruism, and selfishness.[6]

While humans have evolved in groups that have competed with each other, from family to clan to religion to nation, is it not possible that one more evolutionary step can be taken? Competition subsumed within the context of cooperation is constructive, while cooperation subsumed within an overall context of competition is the world system we have inherited, and it is destructive. We will remain primitive until we reverse the order of primacy. Fortunately, we have begun the evolution—in international institutions, communication, and education—of this new way of relating to each other.[7] We just have to speed it up.

THE FIXED STAR: THE INVARIANCE IN RELATIVITY

World War I shocked many into the need for cooperation to replace competition. In 1919, Romain Rolland wrote a *Declaration of the Independence of the Mind*, signed by Albert Einstein, Jane Addams, Bertrand Russell, and other luminaries, which stated in part:

> Our role, our duty is to maintain a fixed point, to show the pole star amidst the storm of passions in the darkness. Among these passions of pride and mutual destructions, we do not single out any one, we reject them all. We commit ourselves never to serve anything but the free Truth that has no frontiers and no limits and is without prejudice against races or castes. Of course, we do not dissociate ourselves from Humanity. We toil for it—but for all humanity. We do not recognize peoples—we acknowledge the People—unique and universal.[8]

The impacts of climate change, the plunging numbers of insects, spreading desertification, these and other planetary calamities are not dissimilar to the disruption of world war, and similarly force us to think about the big picture, and to see the fixed star of planetary health. The truth of our time is that this is necessary. What once seemed impractically idealistic—universal identity—is now the only practical viewpoint.

So often our urgency to get something done leads us to skip the important step of getting the full scope. Some also may feel that when the lens

6. *The Science of Good and Evil* (2004), 55; Shermer is the publisher of *Skeptic Magazine*.
7. While the twentieth century saw the most violent events in human history, it also saw the explosive growth of international cooperation, from the League of Nations to the UN's seventeen international agencies, see https://ask.un.org/faq/140935.
8. https://www.panarchy.org/rolland/spirit.html.

widens enough to include everybody, or everything, we lose focus. But we gain a larger capacity to focus and to integrate particulars into an inclusive view.[9] This is where we have to start, and end, when we are thinking as a group, and if we are to have a chance of grasping the full extent of our common problems and the full array of options we have for action. This is also why each individual person and viewpoint is important, because each of us will then have to choose what works best for us and how we can contribute.

Universality—the perspective that embraces all—need not limit or cancel or blur or diminish anything except pretense to universality. Because it is simply the totality of things and their relationships, it must affirm rather than erase individuality by viewing each person as contributory to the whole. The sham universality of dictators pretending to represent the whole can be shattered by the recognition of its failure to respect the individual.

That there is a true universal—as exemplified by our shared vulnerability before climate change and other threats to life-sustaining natural systems—is reflected by modern science, which has been misinterpreted. As Einstein scholar Gerald Holton wrote, concerning the "widespread misunderstanding that physical relativity meant that all frameworks, points of view, narrators, fragments of plot, or thematic elements are created equal, that each of the polyphonic reports and contrasting perceptions is as valid or expedient as any other":

> if anything, twentieth-century relativistic physics has taught the contrary: that under certain conditions we can extract from different reports, or even from the report originating in one frame properly identified, all the laws of physics, each applicable in any framework, each having therefore an invariant meaning, one that does not depend on the accident of which frame one inhabits. It is for this reason that, by comparison with classical physics, modern relativity is simple, universal, and one may even say, "absolute." The cliché became, erroneously, that "everything is relative," whereas the whole point is that out of the vast flux one can distill the very opposite: "some things are invariant."[10]

9. *The Whole Earth Catalog*, first published in 1968, featured one of the first pictures of our planet taken from space. That widened focus corresponded with an intimate, personal focus that viewed individuals as shapers of their environment, and its purpose as providing tools for self-education for planet-responsible living: "remotely done power and glory—as via government, big business, formal education, church—has succeeded to the point where gross obscure actual gains. In response to this dilemma and to these gains a realm of intimate, personal power is developing—power of the individual to conduct his own education, find his own inspiration, shape his own environment" All the catalogs are available on the internet archive at https://archive.org/details/wholeearth.

10. "The Shaping of Our Imagination," introductory essay to *Albert Einstein, Historical and Cultural Perspec*tives, xv.

Affirming the fact of universality—understanding that relative values, positions, and perspectives are contained within a shared context (such as the invariance of the speed of light in every frame), is resonant with the concept of equality before the law, and the culmination of where inherent, inalienable rights leads. But can there be hope for widespread belief in such unity, given the history of division and war we have inherited?

CULTURAL DEVELOPMENT

In 1932, at the behest of the International Institute of Intellectual Cooperation of the League of Nations, Einstein wrote an open letter to Sigmund Freud to enlist his advice about "delivering mankind from the menace of war." The dialogue that ensued is famous for Freud's initial suggestion that humanity has some kind of death wish. While he did suggest this as a possibility, he also noted that advances had been made in the past, such as when the establishment of the Olympic Games served to unite ancient Greeks, and the spread of Christendom united many diverse European peoples. Though neither held, he observed that both he and Einstein were examples of modern people who hated violence and war.

> The psychic changes which accompany this process of cultural change are striking, and not to be gainsaid. They consist in the progressive rejection of instinctive ends and a scaling down of instinctive reactions.

Answering the question of how long we have to wait until others become pacifists, with a "constitutional intolerance" for war, he opined that

> perhaps our hope that these two factors—man's cultural disposition and a well-founded dread of the form that future wars will take—may serve to put an end to war in the near future, is not chimerical. But by what ways or byways this will come about, we cannot guess. Meanwhile we may rest on the assurance that whatever makes for cultural development is working also against war.[11]

The prospect of environmental destruction may not loom in everyone's mind in the same way as war does, with the immediacy of its horror, but it is certainly analogous for anyone who fully faces the prospects of devastation it brings. Freud nodded toward the evolutionary step we can take.

11. http://www.public.asu.edu/~jmlynch/273/documents/FreudEinstein.pdf.

SOCIALIZING STATES

Since that moment in history, the development of international organizations and communication has exploded to the point that it can be said that a world culture is in development, perhaps best symbolized by the agreement by the UN on Sustainable Development Goals and the beginning of incorporation of its aims into world finance, which is a response analogous to the movement that eventually formed the United Nations. The international institutions may all too often allow powerful states to "simply buy off the cooperation of smaller ones," and the policies of international financial institutions "to reflect the geopolitical interests of the major powers, and the United States in particular," (the realist approach), but international engagement also enhances the capacity of "states to overcome problems of collective action, high transactions costs, and information deficits or asymmetries" (the rational functionalism approach). A third approach, social constructivism, emphasizes and examines how interaction through nongovernmental organizations as well as governmental institutions, in an intersubjective context recognizing universality, can generate a constitutive process that imbues international efforts with meaning, enhancing the possibility that states can become socialized through international cooperation.[12] Although some see a watering-down process, exemplified in the perspective that projections of the Intergovernmental Panel on Climate Change are more conservative than the science would dictate, because of the moderating influences of the participating scientists' employers, the long view of growth in international cooperation indicates the possibility of hope for a world still trapped in primitive aggression and assertion to develop a more mature interactive outlook.

The prospects for hope would be enhanced if the evolution is toward Walt Whitman's idea of democracy as reflecting "the individual en masse," rather than realpolitik balances of power by dominating forces. In a society that faces and turns from the divisive identifications that ignore our actual reality as members of a shared biosphere, what each individual brings to the shared task of preserving it can be combined synergistically. The result of valuing individuals in collaborative combination raises the possibility of benefits far beyond what competition produces—in which only a few winners have the chance to fulfill their potential. In a world of common effort that welcomes the contributions of each individual's talents and energy, we have a chance

12. Lisa Martin and Beth Simmons, "International Organizations and Institutions," in *Handbook of International Relations* (2012), posted at https://scholar.harvard.edu/files/bsimmons/files/ch_13_-_international_os_and_is.pdf.

to produce quality "livingry" (that which assists the act and tasks of living)[13] that exceeds anything produced before in manifesting human potential. The "singularity"[14] that tech enthusiasts point to, at which capability takes off, usually referring to artificial intelligence, should really be applied to what humanity could accomplish if we learn to cooperate instead of struggle against each other. The body of humanity, viewed as a superorganism, needs each individual to be functioning to their potential, as a body needs each cell and organism to function well for it to reach its potential.

EXPANDING OUR PROBLEM-SOLVING CAPACITY AND SCOPE

For years I have been teaching a class at Boston University called the Development of Sustainable Environmental Responsibility,[15] in which we look at the questions involved in getting to where we need to be, from technical to psychological through economic, legal, and cultural, using any useful lens to jointly problem-solve. We begin the semester with a discussion of all the issues we could be looking at. What concerns you? If we were mustering an effort by the new administration to deal with environmental issues, what would you recommend?

Such a question results in lists like this:

- greenhouse gases, flooding, warming seas
- habitat destruction, species extinction
- safe drinking water and adequate supply for the future
- waste and wastewaters from people and industry
- cruel treatment of animals at factory farms
- food deserts in the inner city and policies that favor production of poor nutrition
- industrialization of agriculture
- toxics in products, like PFCs, flame retardants

13. A term invented by Buckminster Fuller. See the 1983 article "Humanity's Critical Path: From Weaponry to Livingry," which states that humanity can evolve away from self-destruction toward world prosperity and peace, https://fullerfuture.files.wordpress.com/2013/01/buckminster fuller-fromweaponrytolivingry1.pdf.

14. "The term singularity describes the moment when a civilization changes so much that its rules and technologies are incomprehensible to previous generations. Think of it as a point-of-no-return in history. Most thinkers believe the singularity will be jump-started by extremely rapid technological and scientific changes," Annalee Newitz, in *Gizmodo*, May 10, 2010, https://io9.gizmodo.com/what-is-the-singularity-and-will-you-live-to-see-it-5534848.

15. https://www.trunity.com/trubook-developing-sustainable-environmental-responsibility-richard-reibstein.html (2018), is the text.

- contaminated sites still not cleaned up
- air pollution, particularly particulates
- environmental justice, disparate impact on the already disadvantaged
- radiation, ionizing and nonionizing
- noise
- nanotechnology
- genetic modification
- overpopulation

Usually, they are a lot longer, but you get the idea. Listing just sixteen is overwhelming. But if you have a class of sixteen people, each can take one and report back. Using a problem-solving approach in which you study the problem in order to understand your options for reducing risks, each student brings back to the team a way to think about the problem that allows us to envision a way back from the brink that we seem to be approaching. When you have a full class taking on a range of issues and finding answers, exploring things we can do about each one, it is extremely helpful. The capacity of a group to teach itself is higher than what I, the teacher, can do alone.

Our system of environmental protection must be ambitious enough to address the entirety of the biosphere and everything living in it. Our system of laws must set as a goal protection of the functioning of the biosphere, the mechanics of life itself. If it does not set this as its goal, we risk failing in all our lesser efforts. We can generate a list of environmental problems and try to solve them one by one, but if we don't work from our actual context of the health of the systems in which we live, we may not even know we are failing until it is too late.

For this reason, many argue that we need to orient environmental protection goals on what we know about the planet's ecological health. As the University of Auckland's Klaus Bosselman wrote in 2010,

> there is a huge and ever-widening gap between the promise of environmental protection and ecological realities. Singular successes are vastly outdone by system failures. . . . In essence, environmental law is hampered by a reductionist approach to its subject, i.e., the environment or more precisely, the human-nature relationship. This relationship is misconceived because of the domination of certain philosophical and cultural traditions in European history. As a consequence, modern legislation to protect the natural environment has developed in a compartmentalized, fragmented, economistic, and anthropocentric manner. For environmental legislation to become effective, broader coverage and better enforcement are not enough. The inherent design flaw in these laws is the absence of a fundamental rule prohibiting harm to the integrity of ecosystems.

> Such a rule requires the acceptance of sustainability as an overarching ethical and legal principle.[16]

Getting the big picture was the concept of the Council on Environmental Quality (CEQ), founded in 1969 as part of the National Environmental Policy Act, in order to assist the president in reporting annually to Congress on

> (1) the status and condition of the major natural, manmade, or altered environmental classes of the Nation, including, but not limited to, the air, the aquatic, including marine, estuarine, and fresh water, and the terrestrial environment, including, but not limited to, the forest, dryland, wetland, range, urban, suburban and rural environment; (2) current and foreseeable trends in the quality, management and utilization of such environments and the effects of those trends on the social, economic, and other requirements of the Nation; (3) the adequacy of available natural resources for fulfilling human and economic requirements of the Nation in the light of expected population pressures; (4) a review of the programs and activities (including regulatory activities) of the Federal Government, the State and local governments, and nongovernmental entities or individuals with particular reference to their effect on the environment and on the conservation, development and utilization of natural resources; and (5) a program for remedying the deficiencies of existing programs and activities, together with recommendations for legislation.[17]

DEVOLUTION

These reports were issued every year until 1997. But the 1995 Federal Reports Elimination and Sunset Act (Public Law 104-66) had the goal of reducing paperwork in government, and House Document 103-7 listed CEQ's reports as something that could be eliminated.[18] To save some small expense, we jettisoned the ability to put before the public the big picture of what is happening to our world. This fits the definition of devolution of societal development in our time.

If the motivation of the House was truly to save paperwork, their act was astoundingly foolish, by any measure of value or comparing benefits to costs. If the motivation was actually to prevent the public from knowing about the true state of the environment, the act can be considered ignoble, a violation of trust by those elected to represent the people, who chose instead to represent only those who profit from continuing the activities that harm our world.

---

16. "Losing the Forest for the Trees: Environmental Reductionism in the Law," *Sustainability* 2, no. 8 (2010): 2424–48, doi:10.3390/su2082424.
17. Sec. 201 [42 USC § 4341].
18. https://ceq.doe.gov/ceq-reports/annual_environmental_quality_reports.html.

Most important, is whether we today repudiate the strategy of burying our heads in the sand.

Although getting the big picture can be said to have been included in the original aspiration of early environmental law, our goals have been reduced in practice, so that basing environmental protection on ecosystem protection has been referred to as recently as 2014 as a "new paradigm." Described as a way of transcending "the immediate management objective (the particular fish stock; the particular endangered species, the particular pollution source, etc.)," it "shifts focus, and recognizes how each individual component is linked to a wide array of intertemporal and cross-scale ecosystem processes."[19] The approach at least theoretically enables accounting for cumulative impacts, and provides greater justification for precautionary actions. Because of its enormous complexity, and the changing state of knowledge, an "adaptive" management approach is necessitated, with constant revisions. This can produce an unstable regulatory environment, inimical to investment planning. This can be mitigated with assistance, anticipatory action, and a business culture that understands the necessity.

In general, businesses can absorb environmental costs and pass them on to consumers, but have difficulty adapting when the rules are constantly changing. But there is a greater obstacle than the resistance of businesses to *a necessarily dynamic* system of law. We have to change the basic orientation of our thinking, and the laws and customs that we have inherited, such as the willful blindness of ceasing to monitor the big picture.

LEGACY BAGGAGE

In order to reshape law so that it effectively addresses the problem of ecosystem protection the historical development of law must be recognized as having developed in service to "an elite, propertied class of white, European males." As Anna Grear of Cardiff University points out, the "logic of law is politically and ideologically conditioned,"[20] and the

> liberal project (of law) could reconcile promises of universality with exclusions in practice. . . . Rights and benefits were linked to the capacity to reason, and

19. Vito De Lucia, "Competing Narratives and Complex Genealogies: The Ecosystem Approach in International Environmental Law," *Journal of Environmental Law* 27 (2015): 91–117, doi: 10.1093/jel/equ031.

20. Citing Alan Norrie's 2001 *Crime, Reason, and History: A Critical Introduction to Criminal Law*.

> the capacity to reason was tied to notions of biological determinism, racial and religious superiority, and civilizational maturity.[21]

Grear comments that "it is particularly revealing that the emergence of a fully capitalistic legal subject in the form of the corporation was pivotal to the colonial development of the international legal order and remains a core if not the central feature of the contemporary global order." Her contention is that the *anthropos* of the human-determined geological age we are in, the Anthropocene, does not represent humanity as a whole, but that our legal system evolved to protect *the propertied class*. Humanity as a whole does not bear the same responsibility for the assault on the biosphere. Our conceptual understanding of the purposes of law require reformulation if the planet is to be saved.

One example of the restructuring of law that is required is re-examination of the personhood of the corporation, the recognition of which at law "reflects and optimizes . . . exclusionary dynamics." The "Law's dominant construction of legal personhood—and law in general, including international environmental law—are thus unresponsive—at a fundamental level—to the ethical implications of the vulnerable embodied biomateriality of the living order."[22] This line of thinking is presaged by historians examining the property-rights based system of liberal democracy, such as the Beards's 1913 *An Economic Interpretation of the Constitution*, or Thorstein Veblen's writings; and resonates with the work of modern observers such as Barbara Ehrenreich, Robert Reich, or Joseph Stiglitz, to name just a few.[23]

THE SHARED CONDITION OF VULNERABILITY

The elevation of disembodied reasoning is often traced to Descartes, its use for power to Bacon, and its gift of imperatives to Kant, but while reason provides the possibility of unity and universality, it can also be misused to mask injustice. In response to this evolution of disembodied rationalism, and its reflection in law favoring an economics of autonomous actors, the legal philosopher Martha Albertson Fineman, of Emory Law School, has developed

---

21. Anna Grear, "Deconstructing Anthropos: A Critical Legal Reflection on 'Anthropocentric' Law and Anthropocene 'Humanity,'" *Law and Critique*; Dordrecht 26, no. 3 (November 2015): 225–49, DOI:10.1007/s10978-015-9161-0. The quoted excerpt is from Ratna Kapur's 2007 *The Citizen and the Migrant: Postcolonial Anxieties, Law, and the Politics of Exclusion/Inclusion. Theoretical Inquiries in Law.*

22. Ibid., 240–41.

23. For an illuminating treatment see Woody Holton's 2007 *Unruly Americans and the Origins of the Constitution*, about how the relatively stable compromise reflects essential tensions between property and democracy.

an ongoing project on Vulnerability and the Human Condition, to correct our misguided idea of freedom through restraint of government action and ground our collective purpose in real need. The currently dominant concept that "State interference with an assumed meritocracy and a market constructed as 'free' would be in violation of the principles of individual liberty and autonomy and an encroachment on freedom of contract," should be replaced by recognition of our actual status as life-forms vulnerable to many forms of injury. Vulnerability is "the characteristic that positions us in relation to each other as human beings, as well as forming the basis for a claim that the state must be more responsive":

> Society is conceived as a collection of self-interested individuals, each of whom has the capacity to manipulate and manage their independently acquired and overlapping resources. Importantly, rather than being dependent on or asserting entitlement to the provision of socioeconomic goods by the state, the liberal subject demands only the autonomy that will enable him to provide for himself and his family. His demand for liberty is refined as the freedom to make choices, the right to contract. Significantly, this demand for liberty on the part of the individual effectively operates as a restraint on the state, which is deterred from interference with individual liberty, even for the purpose of ensuring greater social equality.
>
> The image of the human being encapsulated in the liberal subject is reductive and fails to reflect the complicated nature of the human condition. A vulnerability analysis asks us (and our economists, philosophers and politicians) to embrace a more complex reality by bringing human dependency and vulnerability back into the center of the inquiry into what it means to be human. A vulnerability approach replaces the liberal subject with the "vulnerable subject." The vulnerable subject is the embodiment of the realization that vulnerability is a universal and constant aspect of the human condition. Dependency and vulnerability are not deviant, but natural and inevitable.[24]

Another aspect of vulnerability theory, as developed by Fineman and her colleague Atieno Mboya Samandari, is that a society that strengthens relationships and has responsive government will be more resilient.[25] An embodied law reflecting our embodied nature would de-privilege the disembodied corporation relative to actual people, and help reorient the concepts underlying our system of law, to guide economic development toward sustainability.

---

24. Martha Albertson Fineman, "Equality, Autonomy, and the Vulnerable Subject in Law and Politics," chapter 1 in Martha Albertson Fineman and Anna Grear, *Vulnerability: Reflections on a New Ethical Foundation for Law and Politics* (2013).

25. https://podcasts.apple.com/us/podcast/interview-with-dr-atieno-mboya-samandari/id1549956076?i=1000505962468.

MORE CAN BE SHARED

Such efforts as The Program on Corporations, Law, and Democracy,[26] Tellus Institute's Program on Corporate Redesign,[27] and such analysts as Jeffrey Sachs[28] and Joseph Stiglitz;[29] along with political movements (with moments such as when left-wing participants in Occupy Wall Street seemed to agree with right-wing participants in the Tea Party, or Trump supporters, pre-2016, also supported Bernie Sanders), indicate a growing consensus for a reorganization of priorities from unmitigated wealth extraction—what corporations do—to a more shared prosperity. Ralph Nader and Carl Mayer argued in a 1988 *New York Times* op-ed for a constitutional presumption favoring people over corporations, observing that

> The corporate drive for constitutional parity with real humans comes at a time when legislatures are awarding these artificial persons superhuman privileges. Besides perpetual life, corporations enjoy limited liability for industrial accidents such as nuclear power disasters, and the use of voluntary bankruptcy and other disappearing acts to dodge financial obligations while remaining in business.
>
> The legal system is thus creating unaccountable Frankensteins that have human powers but are nonetheless constitutionally shielded from much actual and potential law enforcement as well as from accountability to real persons such as workers, consumers and taxpayers.[30]

A critically needed reform of conventional economics is the new school of thought termed "ecological economics." Conventional economics is more accurately termed chrematistics—the study of wealth. Ecological economics is, as described by the University of Vermont's Robert Costanza, a transdisciplinary approach, bridging economics and other disciplines such as psychology, in "an attempt to look at humans embedded in their ecological life-support system, not separate from the environment."[31] As Herman Daly of the University of Maryland wrote in *Scientific American*,

> Relying on growth . . . might be fine if the global economy existed in a void, but it does not. Rather the economy is a subsystem of the finite biosphere that supports it. When the economy's expansion encroaches too much on its

26. https://www.poclad.org/.
27. Led by Allen White, publications at https://www.tellus.org/about/people#.
28. His 2011 *The Price of Civilization*, describes how the US is a "corporatocracy."
29. His 2019 *People, Power, and Profits* locates wealth in the "creativity and productivity" of people, not corporations.
30. https://www.nytimes.com/1988/04/09/opinion/corporations-are-not-persons.html.
31. https://insights.som.yale.edu/insights/what-is-ecological-economics.

surrounding ecosystem, we will begin to sacrifice natural capital (such as fish, minerals and fossil fuels) that is worth more than the man-made capital (such as roads, factories and appliances) added by the growth. We will then have what I call uneconomic growth, producing "bads" faster than goods—making us poorer, not richer.[32]

REAL WEALTH

One approach is to include accounting for natural resources in national economic assessments. The National Research Council recommended this approach in its report, *Assigning Economic Value to Natural Resources*, finding that

> national income accounts, which, in their calculations of GNP or GDP (Gross National or Domestic Product, the conventional assessment of economic health used by governments throughout the world), make no or very inadequate allowance for natural resource use and environmental degradation. As a consequence, there is concern that national policy underprotects natural assets, leading to their excessive depletion and degradation, relative to what would be a more balanced national strategy.[33]

New methods of accounting include the Genuine Progress Indicator, which "incorporates the environmental costs of pollution and climate change as well as the social costs of inequality, homelessness, debt and family breakdown."[34] In a piece posted by the World Economic Forum in 2018, Pooran Desai, CEO of OnePlanet.com, referred to the GPI as a "Quality Adjusted" GDP and described how "Quality Adjusted Dollars" would bring about recognition that

> not every dollar transaction has the same value or quality. A dollar spent on renewable energy invests in the future, while a dollar spent on fossil fuel energy depletes it. Thus, conceptually we can, as economists, say "integrate the externalities"—but crucially at the point of transaction, rewarding good transactions in such a way that our brains are stimulated to do good.[35]

---

32. "Economics in a Full World," September 2005, https://steadystate.org/wp-content/uploads/Daly_SciAmerican_FullWorldEconomics(1).pdf.
33. National Academy Press (1994), https://www.nap.edu/download/4844, introduction.
34. https://sustainable-economy.org/new-measures-progress/.
35. https://www.weforum.org/agenda/2018/05/gdp-is-destroying-the-planet-heres-an-alternative/.

THE PERCEIVED THREAT TO FREEDOM

These concepts about reorganization of law and economics and the underlying conceptual foundations of what we have inherited must gain wide currency and acceptance in order to have the effect of bringing about the fundamental transitions in systems that we need. The public must be helped not to misinterpret the drive for beneficial reform as endangering lifestyles they prize. A vision of a new economic system that will provide modern conveniences and comforts is necessary. When these ideas of reform remain abstract, they are vulnerable to manipulation, as when ideas of humane or responsible capitalism are said to be masks for a secret totalitarian socialist or communist plot, in actuality motivated by a desire to end private ownership.[36] For example, the American Society for the Defense of Tradition, Family, and Property website contains a 2019 piece entitled "Green is the New Red," which contains the statement that "Vaclav Klaus, the first president of the Czech Republic after the end of the Soviet dictatorship, warned that 'Celebrating the end of communism is inappropriate. It is creeping back in different forms under different flags and slogans.'" The author continues, "As someone who lived under communism for most of my life, I feel obliged to say that the biggest threat to freedom, democracy, the market economy and prosperity at the beginning of the twenty-first century, is not communism or its various softer versions. It was replaced by ambitious environmentalism."[37] Václav Klaus was in fact the second president of the Czech Republic. *Václav Havel* was the first, and *he* said:

> Today's world, as we all know, is faced with multiple threats. From whichever angle I look at this menace, I always come to the conclusion that salvation can only come through a profound awakening of man to his own personal responsibility, which is at the same time a global responsibility. Thus, the only way to save our world, as I see it, lies in a democracy that recalls its ancient Greek roots: democracy based on an integral human personality personally answering for the fate of the community.[38]

---

36. Representative Chris Stewart (R-Utah) founded an Anti-Socialist Caucus after Representative Alexandria Ocasio-Cortez outlined a vision of a Green New Deal. Dan Zak, "A Green New Deal Ignites an Old Red Scare," *Washington Post*, May 8, 2019, Style Section.

37. https://www.tfp.org/green-is-the-new-red-the-metamorphosis-of-communism/. For just two of many other examples, see https://www.politicususa.com/2013/02/15/rush-limbaugh-claims-environmentalists-coming-steal-electricity.html; https://www.breitbart.com/europe/2020/12/13/report-western-environmentalists-are-the-chinese-communists-useful-idiots/.

38. Onassis Prize for Man and Mankind (1993), https://quotepark.com/authors/vaclav-havel/.

THE VALUE OF THE REGULATED MARKET

It is possible for a dictatorial government to redefine the role of the corporation and to reshape the economy so that natural assets are preserved as healthy ecosystems for the benefit of all. But the history of the world largely bears out Lord Acton's dictum that power corrupts, and absolute power corrupts absolutely. The laissez-faire approach of leaving the organization of our lives to the market may seem like a recipe for freedom, but it produces monopolistic and oligarchic arrangements as actors gain dominant market share, unless there is intelligent regulation. An unregulated market is analogous to when the teachers stay inside and the bullies take over the schoolyard. Competition must be subsumed within a caretaking context. Even Adam Smith believed in the importance of governance. In the *Theory of Moral Sentiments*, he wrote that a government official is

> entrusted with the power not only of . . . restraining injustice, but of promoting the prosperity of the commonwealth, by establishing good discipline, and by discouraging every sort of vice and impropriety; he may prescribe rules, therefore, which not only prohibit mutual injuries among fellow-citizens, but command mutual good offices to a certain degree.[39]

This is hardly laissez-faire, though he is hailed by so many as the philosopher of hands-off governance. It is hard to see him agreeing with Friedrich Hayek that regulation of the market is the "Road to Serfdom." The very concept of the "market failure," often used to refer to the ability of commercial actors to impose hazards and damage on others, or on the environment, with impunity, is itself a failure to understand the place of the market—it is a product of the misbegotten idea that societies should be organized by markets, rather than that markets must serve societies. The market is a useful concept and is indeed a manifestation and necessary component of freedom. But without some external control, it devolves to the domination of powerful players. Governance is the necessary factor to ensure that common goods are not despoiled and the market remains uncaptured. Markets value that which is traded and we must value markets for the freedom in exchange that they provide. But they fall far from providing all of society with what all of society needs.

And that which must be left alone is not valued by the market. The market only values exchange, and not existence unless something is preserved for exchange later. Its basic unit, money, only represents and does not embody

39. Adam Smith, *Theory of Moral Sentiments*, 81.

the inherent value of life or the substances and conditions essential to it. We require social organization that values existence.

Besides market ideologists and libertarians, opponents of governance are sometimes found amongst those who value community, such as Peter Kropotkin, a Russian born into nobility who, in 1872, declared himself an anarchist, focusing on the capacity of people to self-organize and self-protect. In the extreme environment of Siberia in winter, he

> failed to find—although I was eagerly looking for it—that bitter struggle for the means of existence, among animals belonging to the same species, which was considered by most Darwinists (though not always by Darwin himself) as the dominant characteristic of struggle for life, and the main factor of evolution.[40]

Instead, he found "mutual aid" as a key factor in evolutionary success, and though "there are a number of evolutionists who may not refuse to admit the importance of mutual aid among animals," they "will refuse to admit it for Man. For primitive Man—they maintain—war of each against all was the law of life." Kropotkin thought government just got in the way of natural human tendencies to cooperate. We can hope he's right, but use government until such time as that wonderful behavior dominates.

The belief in unrestrained struggle of each against all shaped American law during the Gilded Age of the late 1800s, when many judges imagined a right to freely contract by laborers who actually had no choice but to accept what employers offered. It was not until Justice Oliver Wendell Holmes wrote, in 1905, that "The Fourteenth Amendment does not enact Mr. Herbert Spencer's *Social Statics*" that this began to change. Spencer was a leading social Darwinist, articulating the theory that those who succeeded in life were those most fit to succeed and this was as it should be. This theory justified the laissez-faire philosophy that led judges to overturn laws that protected workers, women, and children.[41]

Scientists today, like Peter Kropotkin, point to the history of successful human organization in small groups.[42] But we now live in a world too crowded for that to be our only method of self-organizing. Democracy

---

40. *Mutual Aid, A Factor of Evolution* (1902), https://ecology.iww.org/PDF/Kropotkin/Mutual%20Aid.pdf?bot_test=1.

41. In *Lochner v. US*, 199 US 45, Holmes wrote that "a constitution is not intended to embody a particular economic theory, whether of paternalism and the organic relation of the citizen to the State or of laissez faire." He dissented from a decision that overturned a New York law limiting the hours a bakery could force its employees to work. The majority found that this state legislation interfered with the right to contract, and that this right was guaranteed by the Fourteenth Amendment (which was actually enacted to protect the rights of newly freed slaves).

42. Robin Dunbar and Richard Sosis, "Optimising Human Community Sizes," *Evolution and Human Behavior* 39, no. 1 (January 2018): 106–11, https://www.sciencedirect.com/science/article/pii/S109051381730209X.

remains the only feasible choice. James Madison theorized in the *Federalist Papers* (Number 10) that a representative democracy could succeed in holding together a large population because it would

> take in a greater variety of parties and interests; (making it) less probable that a majority of the whole will have a common motive to invade the rights of other citizens; or if such a common motive exists, it will be more difficult for all who feel it to discover their own strength, and to act in unison with each other.

Shermer quotes E. O. Wilson:

> Hunter-gatherers are egalitarian, not because they lack selfish impulses, but because selfish impulses are effectively controlled by other members of the group. In human hunter-gatherer groups, an individual who attempts to dominate others is likely to encounter the combined resistance of the rest of the group.[43]

While it is possible to say that the current situation of fierce partisanship and in particular the focus of distrust in government disproves Madison's theory that the same dynamic of group interest can hold in a very large society as holds in a small one, it is also possible to say that the experiment in democracy that is the American story has yet held together and has managed to invent modern environmental law. That there has been a destructive setback, epitomized but not limited to Trump, does not negate but rather illuminates the force of this progress—it threatens narrow, proprietary interests. That a key strategy of those opposing environmental progress has been to sow distrust in democracy shows that Havel was right. If we heed his words, we can build trust in our ability to recognize our common fate, and use the tools of democracy to forge methods of governance that truly protect. No cavalry, no hero, no particular group, will likely arrive to save us, but there is a possibility that with more clarity and communication of the realities we face and the tools for salvation available to us, that we can summon the spirit of cooperation, which can. What other avenue presents itself, but to continue evolution until the interest of the group asserts itself over the will to dominate that some individuals possess?

If such a political consensus can be built, we can envision what a sensible world democracy could produce:

The science necessary to know when ecosystems are endangered
The technologies and practices that can mitigate the dangers

---

43. *Darwin's Cathedral*, 21; Shermer, 55.

The economic systems that can share prosperity
The social organization that allows for the flourishing of human potential.[44]

FREEDOM OF CHOICE, FREEDOM TO HAVE GOVERNMENT

Such a government, valuing freedom within a shared context, would seek to encourage the spontaneous development of initiatives to meet needs through free enterprise, asserting controls to prevent domination of markets, which limits participation and creative contribution. But it would also understand needs and desires, both common and individual, valuing all, and acting to inspire and cultivate constructive responses where the market did not produce them. It would not be accurate to say that this was government acting to correct market failures, for it was incorrect to think that the market would provide all of what humanity needs. It would be a government that supports markets that support and enhance existence.

Freedom of choice requires markets, and they have been shown to be more efficient in many ways than central planning, as well as engines for stimulating innovation and creativity, rewarding enterprise and energy to produce. Therefore, it seems clear that they must be valued and preserved for what they do. But we must also disabuse ourselves of the unjustified belief in their divine ordination and abandon, as did Justice Holmes more than a century ago, bowing to the right of the people to enact laws to protect the public interest.

Those who produce goods are assuredly of benefit, and we can assume most deserve the profit they gain. Corporations are not bad per se. But no corporation deserves what no individual can lay claim to—the right to harm. The assumption that our lives are best left to their discretion is unwarranted. The idea of getting government off the backs of industry is the product of atrophy of the moral sentiments the high priest of capitalism himself expounded.

When he was Dean of the Yale School of Forestry and Environmental Studies, former head of the Council on Environmental Quality and founder of the World Resources Institute Gus Speth told the *Hartford Courant* that

> The utopian fantasy is to think that we can keep doing what we are doing today.

44. See, for example, *Flourishing: A Frank Conversation about Sustainability* (Stanford Business Books, 2013), by John Ehrenfeld and Andrew Hoffman: "this book tells a new story, driven by being and caring, as opposed to having and needing, rooted in the beauty of complexity and arguing for the transformative cultural shift that we can make based on our collective wisdom and lived experiences," https://www.johnehrenfeld.com/johns-books/.

> Our present rates of economic growth, consumption, population growth and waste production will quickly deplete the Earth's resources. . . . What is needed . . . is a radical change in the economic system that takes into account the environmental costs of doing business and refocuses society on building more sustainable ways of living. This will require a change in how we conduct politics, and, more important, a change in consciousness and rethinking the values by which we lead our daily lives.[45]

We may presume that it will be hard. We may imagine that it could be surprisingly easy. (After all, not having to worry about environmental catastrophe is desirable). How we change is a big question to tackle. But about one thing there can be no doubt: we must. Just because it's a big thing is not an excuse not to get going.

A moral discourse is necessary to work out the foundations of the systems of governance we ought to have.

It is important to see the moral content, or lack of it, in how we make environmental and public health decisions today.

GO/NO GO QUESTION OF ACCEPTABLE RISK INSTEAD OF RISK REDUCTION RESPONSIBILITY

To reconstruct the regulatory system and make it work better we need to remove the chains that have been placed on rule-making. The big picture tells us we are stalled and need to remove obstacles to progress. That involves having a concept that is big enough to provide forward movement. One such concept is the moral approach to risk reduction. The moral approach searches for universal ethical concepts and applies them in the design of government's response.

Take Problem X: we ask, how bad is the risk? Asking whether the risk is "acceptable" before deciding to act is a morally primitive stance. It suffers from the hubris of assuming risk can be well enough determined, and it transposes the right of acceptance from the victim to someone with questionable authority to decide for them.

Compare it to the more advanced approach of adopting ALARA as a universal principle. ALARA is As Low as Reasonably Achievable, and we apply it to nuclear operations. These are considered especially dangerous and frightening because radiation has potentially gene-altering impacts that can last through the generations and humanity has already increased background radiation above what we would have experienced naturally. We know we

45. David Funkhauser, "Yale Dean is Idea Man of Climate Change," April 17, 2008, https://www.courant.com/news/connecticut/hc-xpm-2008-04-17-ctspeth0417-art-story.html.

need to just get that exposure down as low as we can. We can apply the same principle to all other risks, even though we can be far more casual about them. There's a big difference between the approach of continuously trying to reduce the level of insult and the approach that says, tell me how much harm can be accepted before we decide we need to do *anything*.

We should always do something if there is a chance that we will harm others, even with risks that are only perceived.[46] If people don't feel safe, even though there is no actual threat, the fact that they don't feel safe is of concern. We should always take *some* action. To question whether we should take action is essentially uncaring. It is the question of what action to take that is before us, and which is relational, moral, and leads to potential improvements.

The resistance to regulation has forced us to adopt the amoral stance and sometimes even when risks are quite perceptible, advocates must work very hard to prove that they should be regarded as unacceptable. We have essentially adopted the procedure of assuming chemicals and many business operations are innocent until proven guilty, a standard that allows harm until it becomes undeniable.

It may be that the resistance to recognizing risk is related to the fear that excessive costs or penalties will result. That fear should be faced head-on, and not by denying the existence of risk or—as is often the case—denying that it is being imposed, typically without knowledge or consent.

Displacing risk reduction action from the primary position can be understood as a violation of rights and values. It may be ignoring whether someone has the right to impose a risk on another person. Now it may be that in the final analysis, when we have balanced everyone's perspectives, we do not want to shut down an existing factory because it will hurt the local economy. But that doesn't mean that it can then pollute its neighbors with impunity. Using a rights and values approach, we work to try to make everyone better off. All polluting sources should be under some continuous pressure to reduce pollution as they have no inherent right to harm their neighbors, despite that we have given them a "permit." If the factory does not know how to reduce its impacts, government should help them. But the process should not be an action/no action decision, based on a technocratic evaluation of risks. The awarding of the permit should be considered a temporary measure. There are two simple reasons why we should dispense with risk assessment as a go/

46. In many forums MIT's Nicholas Ashford has promoted the idea of technology options analysis as a main tool of regulatory response to risk. In a 2009 conference on new ideas in risk regulation, hosted by Adam Finkel of the Penn Program on Regulation, he pointed out the old paradigm that risk management begins after risk assessment could be discarded in favor of one in which risk management starts the process and risk assessment is then harnessed to evaluate solutions, https://media.rff.org/archive/files/sharepoint/Documents/Events/090622_Risk_Regulation/090622_Finkel.pdf.

no go decision point for action: it is undemocratic, but it is also hubristic—it assumes we can know all we need to know in order to assess risk, but we can't.

LET COST-BENEFIT ANALYSIS INFORM HOW WE PROCEED, NOT WHETHER WE ACT

If, under the present system, we have established that there is a significant risk, then we proceed to evaluating the costs and benefits of taking action. Just like with understanding risk, there is nothing wrong and it's all to the good to better understand costs and benefits, but there is everything wrong with deferring to a cost-benefit analysis (CBA) to decide if any action is necessary. As with risk assessment, it dispenses with a democratic process of considering rights and values. For comparison, think about the Clean Air Act Amendments of 1970.

On the floor of the Senate, September 21, 1970, principal sponsor Senator Muskie of Maine said:

> The first responsibility of Congress is not the making of technological or economic judgments—or even to be limited by what is or appears to be technologically or economically feasible. Our responsibility is to establish what the public interest requires to protect the health of persons.
>
> This may mean that people and industries will be asked to do what seems to be impossible at the present time. But if health is to be protected, these challenges must be met. I am convinced they can be met. . . . Our responsibility in Congress is to say that the requirements of this bill are what the health of the Nation requires, and to challenge polluters to meet them.[47]

This was challenged again and again but Congress was clear, in this example, that health came first. The courts have upheld that decision.[48]

In 1980, corporations putting lead into commerce claimed that EPA went beyond its authority by failing to consider the economic impacts of its limits on lead air emissions.[49] But Judge Skelly Wright looked to the intent of

47. http://abacus.bates.edu/muskie-archives/ajcr/1970/CAA%20ESM%20Opening.shtml.

48. See, for example, *Union Elec. Co. v. EPA*, 427 US 246 (1976); where Judge Marshall writes for the Supreme Court that "Congress intended claims of economic and technological infeasibility to be wholly foreign to the Administrator's consideration . . . the 1970 Amendments to the Clean Air Act were a drastic remedy to what was perceived as a serious and otherwise uncheckable problem of air pollution. . . . These requirements are of a 'technology-forcing character,' *Train v. NRDC*, supra, at 91, and are expressly designed to force regulated sources to develop pollution control devices that might at the time appear to be economically or technologically infeasible." This conformed to the national consensus for an environmental future.

49. *Lead Industries Association v. EPA*, 647 F.2d 1130 (1980).

Congress and found that the law talked only about health and not technical or economic feasibility, and that the Senate Committee "determined that existing sources of pollutants either should meet the standard of the law or be closed down, and in addition that new sources should be controlled to the maximum extent possible to prevent atmospheric emissions."

Unfortunately, since then this principle has not been applied to all that many other aspects of risk. We've been cautious about going all out to reduce them. Congress has refrained from again using the strong language used for the "criteria pollutant" section of the Clean Air Act, and while previous presidents had used CBA to understand regulations, Ronald Reagan instituted a strict CBA test for new regulations, essentially instituting the practice of preventing any new regulations that could not prove that benefits exceeded costs.[50] Clinton modified this but did not completely remove it, and CBA still has an important limiting effect on rule-making. In training for EPA managers during the Clinton era, economist Holly Stallworth noted that Reagan's Executive Order

> introduced not only intractable conflict between EPA and OMB, it ushered in a new kind of political economy in regard to environmental issues—one where the environmental quality was treated less as a public good subject to value-based decisions but as a private good where price-based calculations were deemed to best inform. This departed significantly from other public policy issues such as police or fire protection or national defense—where decisions arose from society's values, beliefs, common needs, sense of social justice or risk aversion. Public goods theory developed since the days of Adam Smith long recognized the need for government intervention in a number of select areas plagued by the "free rider" problem—also called the "tragedy of the commons." A classic example would be the fisherman who is unwilling to curtail his harvest from an over-exploited fishery because his private gains from over-fishing run counter to society's long-run interest in maintaining the fishery. Many environmental goods and services meet the test of a public good—indivisibility of ownership or consumption. Therefore private parties, acting on the basis of self-interest, will not have sufficient incentive to pay their fair share or protect a common property resource.[51]

50. "President Reagan's Executive Order 12291 imposed by administrative decree, a decision criteria of maximizing net benefits over costs, regardless of whatever health or technology-based standards were specified in the statute," Holly Stallworth, *The Economics of Sustainability* (US EPA Watershed Academy Web, 1997), 21, http://www.epa.gov/watertrain.

51. Holly Stallworth, "The Use of Economics in Environmental Decision-Making," *Annals of New York Academy of Sciences* (1997), originally published in published in *New Solutions, A Journal of Environmental and Occupational Health Policy* (1996): 8–13. Stallworth was with EPA's Office of Policy, Planning, and Evaluation.

CBA applied as a test for whether a rule should be made ignores this. It also ignores a number of other things. By simply totting up costs on one side and benefits on the other, it dispenses with any question of who benefits and who pays. Let's say a very rich corporation must pay something, but can easily afford to do so, and on the other side we have poor people whose health is affected. These are not commensurable. It's not even adequate to say we are comparing apples and oranges, for at least they are both fruit. It is more like comparing nickels and functioning lungs. Another aspect of reality that is ignored is the priceless quality of health, or life. In the "Rights of Statistical People," Lisa Heinzerling asks a key question:[52]

> The use of cost-benefit analysis to evaluate the wisdom of life-saving regulatory programs presents a puzzle. Deciding to allow one person to harm, even to kill, another person on the basis of how much it costs the person doing the harm to refrain from doing it denies the person harmed a right against harm. It makes a person's freedom from harm, indeed her life, contingent upon the financial profile of the life-threatening activity. The puzzle is that we do not allow this kind of cost-benefit balancing in all life-threatening contexts. We do not, for example, believe that so long as it is worth $10 million to one person to see another person dead, and so long as current estimates of the value of human life are lower than $10 million, it is acceptable for the first person to shoot and kill the second. Indeed, in this setting we refrain entirely from placing a monetary value on life. Yet when it comes to regulatory programs that prevent deaths-deaths also due to the actions of other people-it has become commonplace to argue that the people doing the harm should be allowed to act so long as it would cost more for them to stop doing the harm than the harm is worth in monetary terms. Why are these two situations coming to be viewed so differently?

The answer involves seeing the history of the application of CBA and risk assessment and other methods for short-circuiting what needs to be a values- and rights-based democratic process. We cannot escape making value judgments by deferring to mathematical operations that make them disappear. If we assign an infinite value to anything we blow up the equation. But life has infinite value. Therefore, equations are inadequate and we are forced to work things out as justly as we can.

We need to do what Congress found itself able to do in 1970, and make rules according to our sense of what's right and what's needed.

---

52. *Harvard Environmental Law Review* 24 (2000): 189–207, https://scholarship.law.georgetown.edu/cgi/viewcontent.cgi?article=1322&context=facpub.

THE DISCOUNTING OF FUTURE LIFE

To see the moral blindness of formulaic decision-making consider how future generations are treated. In order to come up with a number representing the value of taking an action today, we have to look at what happens over time. Let's say we want to promulgate a rule that will reduce mercury emissions. If we predict that the result of the rule will be that we will have 10 million tons less mercury pollution, the fact is that this will be a benefit that will last forever. Once we put that mercury out there we can't get it back. It keeps poisoning things. Maybe we will learn that natural processes will sequester it somehow, but until we learn that, we need to assume it will remain biologically available to be taken up again and again. This is true of a great many things we extract from under the earth.

If we try to assess the value of the benefit of removing that mercury, we quickly reach an infinite value if we use a very long time frame. So to keep the formula from blowing up, we limit that time frame. We use a net present value, *discounting the value of something in the future*. This is just a way of not saying we don't recognize a responsibility to the people or animals of the future.

It would be excusable if the wounds we inflict would heal by the time they came around, but for wounds that will persist the calculation is just that old trick of pretending something isn't there.

Recently, an environmental law professor asked if other environmental law professors had something to recommend for students to read to help them understand the concept of discounting the future value of money that is used in cost-benefit accounting. I wrote:

> I like this article because I can get a sense of what all that mathematics is about, and it makes the assumptions clearer.
>
> Maureen L. Cropper in Resources, "How Should Benefits and Costs Be Discounted in an Intergenerational Context?" June 14, 2013, https://www.resources.org/archives/how-should-benefits-and-costs-be-discounted-in-an-intergenerational-context/.
>
> In particular, this makes one thing very clear:
>
>> The consumption rate of discount reflects the rate at which society is willing to trade consumption in the future for consumption today. Basically, we discount the consumption of future generations at a higher rate because we assume future generations will be wealthier than we are and that the utility people receive from consumption declines as their level of consumption increases. To illustrate, if per capita consumption grows

> at 1.3 percent per year, in 200 years it will be over 13 times higher than today. So it makes sense to discount an extra dollar of consumption received 200 years in the future.

> I tell my students that even though there may be a lot of math in an article, they should see if they can't find the assumptions that are used. I ask them if we have the right to assume that our heirs will be better off than we are. The prediction of wealth uses the financial model, which can extend out indefinitely, but that exists within a finite biosphere which is already visibly becoming poorer in resources.

I got two responses—one said, "Point well taken!" And the other said, "It's even dumber than that."

CBA can be unjust in that it equates saving nickels for a rich corporation with an assumed cost for imposing harm on others, and it is unjust in ignoring our responsibility to the future, but it is also ridiculous in its arrogance in thinking that we can calculate the value of the benefits. What is the benefit to you of not being just a little poisoned? Besides the issue of someone else deciding that you can be, do you believe that anyone can know enough to say that the toxic entering your body is nothing to worry about? Have they added up the cumulative effects of all the different toxics entering your body? Do they know exactly what will happen in your body? Do they know all about your particular vulnerabilities or everything about what that toxic substance does? The answer is that they don't. No one does. But we perform the risk and cost assessments assuming that we know enough to produce answers that will be protective. The toxic body burden assessments show this is not working. The oceans show this is not working.

Adopting a little humility would help us to realize that we ought to do rule making differently. If someone is being hurt or is worried about being hurt, that's our start. Then we use risk assessment to understand the risk better, not to decide whether or not we act, but to help us determine what action to take. The moral principle has to be that we always act to reduce the risk—actual or perceived. That sounds tiring but it's simply the case until we transition from a wasteful, polluting society to one that we can be proud to hand down to future people. The moral approach would always apply ALARA, the question only being how, to what extent, with what level of effort. Assuming the necessity of some action is moral, as it is in relationship with the dignity of each potentially harmed individual.

THE ARRAY OF OPTIONS

Applying the moral imperative to try to reduce risks wherever we can, instead of wasting time debating whether to take any action at all, and getting the big picture, leads to options analysis. The next question after perceiving a problem is to ask what are the choices we have for action. Looking at the risks and costs helps us choose our actions. Once we have identified the range of options, we can use risk assessment to estimate how much risk each would reduce, and we can do a cost-effectiveness analysis to estimate how much each action would cost. These are useful tools, if they serve the purpose of action, not if they inhibit it.

The big picture perspective is a necessary part of developing the capacity to bring about a healthy world to live in. The formulaic approaches we have been using are sophisticated math but reductive, crude public policy. They only fit an unreasonable facsimile of reality, and their use can cause us to ignore the moral approach to governance that promises just and stable solutions.

THOUGHT EXPERIMENT

Imagine you are responsible for making a new rule to address a problem. What is the fairest process to use? How can you be as fair as possible to the regulated corporations as to potential recipients of their impacts? How do we best structure the process of rule-making?

Chapter Six

Anticipatory Governance

It is easy to find reference to prevention when people talk about how environmental protection can be improved, as discussed in earlier chapters. Prevention, in the sense used there, is generally applied to existing threats. For example, you are already generating pollution, and the idea is that you ought to look at how you can change what you do so as to eliminate or reduce the problem. But there are also problems that haven't happened yet, and we need to prevent those too. We might use the term "anticipatory governance" to refer to this principle of constructing our system of environmental protection.

Whole new technologies, such as nanotechnology or genetically modified organisms, new chemicals, electromagnetic fields, new products such as artificial turf or new cars with electric batteries, bring new issues with them, and we can do our usual thing of waiting for problems to appear before acting, or we can do our best to anticipate them and head them off. Even such good things as new cars with electric batteries bring new problems—such as the very low rate of battery recycling and the extraction of precious raw materials necessitated by not having closed the circle by recovering all of the batteries.

The term we often hear is "precaution." The precautionary principle, as articulated in the Rio Declaration that came out of the 1992 UN Conference on Environment and Development, included Principle 15:

> In order to protect the environment, the precautionary approach shall be widely applied by States according to their capabilities. Where there are threats of serious or irreversible damage, lack of full scientific certainty shall not be used as a reason for postponing cost-effective measures to prevent environmental degradation.

This makes sense. Anyone who has had a wise older person tell them it is "better to be safe than sorry" knows this principle. Anyone who has thought twice before undertaking a risky activity knows this principle. However, people who think about governance have identified a kind of error that can occur—overregulation. Anyone who has been annoyed by the wise older person trying to keep them from doing something fun because of some possible risk knows about this. As Jose Felix Pinto-Bazurco notes in an article that traces the development of the concept for the International Institute for Sustainable Development, "opponents have decried the potential the principle has for over-regulating or limiting human activity, as we see in the criticism about the establishment of moratoriums on genetically modified organisms (GMOs) in some countries."[1]

But when the risk of harm is very great, even if there is a small chance it will occur, we shouldn't worry too much about overregulation. We may call the mother who has her child put on rubber boots when it's raining overprotective, but if the child has a high fever, we would be very wrong. Even highly respected thinkers have improperly focused on the risk of overregulation. The renowned scientist Freeman Dyson wrote, in 2003, in the *New York Review of Books* that

> the precautionary principle says that when there is any risk of a major disaster, no action should be permitted that increases the risk. If, as often happens, an action promises to bring substantial benefits together with some risk of a major disaster, no balancing of benefits against risks is to be allowed. Any action carrying a risk of major disaster must be prohibited, regardless of the costs of prohibition.[2]

Dyson was at the Institute for Advanced Study, a place for very smart people. But it is not clear where his definition of precaution comes from. With all due respect to someone whose writings on scientific matters have been illuminating, it appears to be a paranoid fantasy. Compare it to what is considered the most definitive articulation, the 1998 Wingspread Statement:

> Where an activity raises threats of harm to the environment or human health, precautionary measures should be taken even if some cause and effect relationships are not fully established scientifically.

---

1. A *Still Only One Earth* policy brief, October 23, 2020, https://www.iisd.org/articles/precautionary-principle. After Peru and Germany established GMO moratoria, critics argued this "would affect the availability of food, especially in developing countries, while proponents defended the cautious approach as an effort to prevent possible negative impacts on biodiversity and the health of people consuming genetically modified products."
2. https://www.nybooks.com/articles/2003/02/13/the-future-needs-us/.

> In this context the proponent of an activity, rather than the public bears the burden of proof.
>
> The process of applying the Precautionary Principle must be open, informed and democratic, and must include potentially affected parties. It must also involve an examination of the full range of alternatives, including no action.[3]

Dyson is hardly the only respected figure to act as if the idea of precaution was itself a threat. Precaution simply makes sense, and fearing it does not. It is true that overregulation occurs and that it is to be avoided. It is not a reason not to try to reduce an impact that must be reduced.

Do we attribute the hysterical reaction to the idea of prevention, to the prospect of regulation of some kind, to government intervention, to an underlying belief that the rights of enterprise, of exploration of the new, must be primary, dominant over other concerns? Certainly, that seems to be the philosophy of many engaged in toxic, hazardous, risky, polluting activities, and our current system of law seems to have evolved to favor economic development by downplaying the possibility of legal liabilities or regulation that would control it.

A HISTORY OF ALLOWING RISKY BUSINESS

For example, the famous English case of *Rylands v Fletcher*, which found Rylands liable for flooding Fletcher's property, after he built a reservoir that burst. The House of Lords found that Rylands was strictly liable, for "the person who for his own purposes brings on his lands and collects and keeps there anything likely to do mischief if it escapes, must keep it in at his peril, and, if he does not do so, is prima facie answerable for all the damage which is the natural consequence of its escape." But this 1868 case has not been followed in the United States, except for "ultrahazardous" activities, because it would interfere with the development of enterprise. As law professor William Wiececk explains, in *The Lost World of Classical Legal Thought*, such liability "imposed the costs of industrial activity without regard to fault, and thus to American judges it was both irrational and redistributive." He describes how Chief Justice Lemuel Shaw "virtually created a whole new domain of law, torts, to structure the increasing volume of litigation occasioned by accidents occurring because of the Industrial Revolution. New doctrines shifted the risks and costs of industrial accidents from employer/producers to employees or consumers, thereby permitting corporations to accumulate

3. At the Global Development Research Center: http://gdrc.org/u-gov/precaution-3.html.

capital for reinvestment." Wiececk quotes Judge Seward Barcolo of the New York Supreme Court in 1852:

> in certain controversies between the weak and the strong—between a humble individual and a gigantic corporation, the sympathies of the human mind naturally, honestly, and generously run to the assistance and support of the feeble and apparently oppressed . . . (but this) is wholly inconsistent with the principles of law and the ends of justice.[4]

Justice then was seen as allowing the economy to develop without government interference. Injustice today is seen when plaintiffs face difficulties getting accountability and compensation for injuries from pollution or other hazards caused by businesses.

We face difficulties passing laws that restrict business activity, or now maintaining the strength of such laws that have already been passed, as illustrated by the fact that the Toxic Substances Control Act, intended as a central effort to control pollution, was reduced to a "dead letter" and only a few chemicals have actually been controlled.

> The Corrosion Proof Fittings case was a landmark event for TSCA. While its legal significance was debatable, the lesson for EPA was that if ten years and thousands of pages of documentation were inadequate to ban asbestos, TSCA's section 6 risk management provision was a dead letter. EPA essentially put regulation pursuant to section 6 on a shelf and spent most of the next two decades seeking voluntary action from industry. Congress, for its part, remained remarkably disinterested in fixing the core provisions of TSCA, adding discrete new titles for asbestos, radon, lead, formaldehyde, and school environments, but leaving the core chemical control provisions untouched.[5]

We know that children as well as others are at serious risk from threats such as neurotoxins, but we have hardly banned the use of them, even though the bans that we have implemented (lead in gas, PCBs) have been enormously successful.[6]

---

4. Page 46, quoting from *Haring v. New-York and Erie R.R.*, 13 Barb. S.C. 2, 15-16 (N.Y. Sup. Ct. 1852).
5. Charles Franklin, "TSCA Reform versus Replacement: Moving Forward in the Chemical Control Debate," *American Bar Association Section on Environment, Energy and Resources,* May 1, 2013.
6. Landrigan and Goldman, "Children's Vulnerability to Toxic Chemicals: A Challenge and Opportunity to Strengthen Health and Environmental Policy," *Health Affairs* 30, no. 5 (2011), http://content.healthaffairs.org/content/early/2011/05/02/hlthaff.2011.0151.full.html.

Perhaps the term "precaution" has been purposely misunderstood. Such semantic warfare occurs.[7] Can those who speak to further the common interest learn to use the words and terms that elicit emotional loyalty, as the propagandists have learned? Can they express meaning that resonates with the public and builds a natural upwelling that overwhelms the manipulations of public feeling?

AVOID BOTH OVER- AND UNDER-REGULATION

Anticipatory governance just means getting ready for what's coming and taking action to clear our way forward. The approach holds out the prospect of political unity because it provides a place for both liberal and conservative perspectives. The liberal who wishes to avoid any new harm risks overregulating, and the conservative who doesn't like government intervention risks under-regulating. Any facilitator, any mediator, any centrist, any independent, any freethinking citizen can see that, if we can get them to the table, we might just find a way to move forward.

We can use anticipatory governance—and the convening of many perspectives to improve the design of it—to create the appropriate sense of urgency concerning what are termed *emerging contaminants.* Too many of these have already emerged. These include:

- nanomaterials
- genetically modified organisms (GMOs)
- endocrine disruptors
- persistent chemicals
- nonionizing electromagnetic radiation

These are all problems that do exist now, but have not yet produced large visible impacts that we cannot politically ignore, so we have not done enough at law to address them. A system of anticipatory governance would take more action. It would not shut down progress but guide it—think of guard rails on mountain roads.

There is a huge response now to the discovery of per- and polyfluoroalkyl substances (PFAS, "forever chemicals") in many sources of drinking water, a growing community of people who don't like the idea of living in

---

7. For a concise overview with examples, see Mark Abadi's "Democrats and Republicans Speak Different Languages—and it Helps Explain Why We're so Divided," *Business Insider*, August 11, 2017, https://www.businessinsider.com/political-language-rhetoric-framing-messaging-lakoff-luntz-2017-8.

strong electromagnetic fields, and there are more and more people aware of and alarmed about endocrine disruptors. Nevertheless, as low as reasonably achievable (ALARA) is not being required. Many good companies are applying precaution. But we have no reason to assume that all of them are and without much regulation, not much real knowledge about what's really happening.

Only small steps have thus far been taken. The frightened response to the perceived risks of nanomaterials and GMOs, that once loomed large in peoples' minds, has faded because no huge damage has become visible, and they have brought us many good things. It should not be doubted that the fears raised about these technologies have led to more careful enterprise. However, the risk that some participants will create new and serious hazards remains, and can be said to have increased because there is more and more use, as the perception of safety increases. This is a paradox for anticipatory governance to see.

Whether already harming us, or not yet harming us, each of these issues presents questions of how we use our system of government, which is based on what our expectations are and should be, of people who place items in commerce. That there might be great benefits to a technology is something to weigh in the balance, but it is irrelevant to the question of what the risk is. This can be a hard thing to communicate but finding out what a risk might be is distinct from the question of benefits. That there may be benefits does not affect the amount or quality of risk. It may affect the choices we make. The important question for policy is whether individuals get to make the choices that affect them, rather than experiencing the risks created by the choices of others.

Note how often a discussion of risks brings the response that there are benefits. The potential or actuality of benefits only means we should work harder on risk reduction when there is risk, so that we can enjoy the benefits without the risks. The presence of benefits is not directly relevant to the question of risk.

We might usefully articulate an ethic that risk creators should be engaged in the risk reduction inquiry at all the times, open about the risks in operations—all aspects—because of a recognition of the consumers and neighbors as stakeholders in the enterprise, affected parties as well as sometimes participants, with rights to know what might affect them. *A big picture context would require producers accept the duty of continuous risk reduction* as part and parcel of operating in a crowded world that is already overloaded with wastes.

Anticipatory governance would also help us avoid overregulation because it gets thought out in advance and not clamped down in a hurry in response to sudden alarm, a familiar pattern in policy-making. It can be used to nurture and cultivate the kind of manufacturing we need.

Genetic modification gone wrong can ruin the gene pool and have terrifying ramifications down the ages. But it can also lead to cures for cancer, sufficient food for billions, species protection or even resurrection. Nanotechnology gives us health advances, too, as well as clean energy and self-cleaning surfaces (so you can eliminate the use of harsh cleaners), but it can also bypass our defenses and spread damaging molecules throughout our bodies.

In the case of endocrine disruptors which can hinder, block or imitate hormones; forever chemicals, which our bodies can't metabolize, and what some call electrosmog (strong electromagnetic fields), the benefits may not be so clear. Do we really need the plastics that use (and leach out) endocrine disruptors? Do we really need stainproof surfaces on our couches, made easy to clean with forever chemicals? What about the ability to download movies with 5G while we walk down the street? Are you so enamored of that possibility that you are willing to overlook the many studies raising concerns?[8]

These are questions about values. When companies put persistent chemicals in products we use, do they have the right to cause the inevitable exposures? When they sold the chemicals for firefighting training, did they not have a responsibility to think about how they would spread like crazy and not break down? We shouldn't calculate risks and benefits unless we have first weighed rights and responsibilities, and the values of our society about what people owe to each other.

PICTURING A DIFFERENT SYSTEM

If we value a fair approach to rights, then we can design a sensible and equitable system of anticipatory government. Imagine how this could work.

DuPont comes to the agency proposing the use of Teflon. It shares what it knows about the chemicals it wants to use. EPA hires independent scientists to evaluate the safety of the chemicals. These scientists exist because firms have grown up to serve the need for independent science, because this function is funded under the Toxic Substances Control Act by a Congress that serves the people. (This is not what happened, but what ought to have happened.) This Congress of our imagination recognizes that the people have a right to know about the safety of chemicals on their frying pans. It also respects the fact

8. "IARC Classifies Radiofrequency Electromagnetic Fields as Possibly Carcinogenic to Humans" (2011), https://www.iarc.who.int/wp-content/uploads/2018/07/pr208_E.pdf, "High Exposure to Radio Frequency Radiation Associated with Cancer in Male Rats" (2018), National Toxicology Program, https://www.niehs.nih.gov/news/newsroom/releases/2018/november1/index.cfm.

that a company needs to be able to compete against other companies, and if it cannot capitalize on what it learns from investing in innovation, then it might not innovate, but people's rights come first.

In our imaginary world, where the system works to protect rights and basic values, and recognizes the rights of business operations to make profits but does not improperly make it primary, the EPA responds that the independent science shows that the chemicals don't break down, thereby creating a risk that, if not completely contained, they will create problems in living things that cannot metabolize them, and so the public has a right to know about this. In our ideal thought experiment, it rules that DuPont may not sell items with it until it has amply demonstrated that use will not lead to release of the chemicals, and certain uses of the chemicals that are dispersive, such as in stain-resistant cloth, and firefighting, are banned, because the releases cannot be controlled, and they must take the items back post-use or somehow ensure proper end-of-life management so that there are no releases then either. And the public must be told about the nature of the chemicals, with warnings and labels and instructions and a deposit refund on how the item should be managed post-use in a system that the company or the company's industry has established and has funded into the future. This is not what happened. What happened was that DuPont hid information from EPA about the toxicity of the chemicals. They were able to do this because TSCA relies too much on the honesty of the company.[9]

We need to do such thought experiments concerning every program that hasn't worked as it should: imagine how it ought to work. We ought to have learned our lesson about persistent chemicals, but we did not. How do we prevent that mistake from happening again? And thought experiments should be applied to programs that are working, to imagine how they could fail.

Recognizing that dispersive uses of PFAS would be disastrous could have led to the training of firefighters with containment, until better firefighting chemicals were developed. Anticipatory thought experiments are the business of government, not assuming that bringing attention to safety will stifle innovation. Requiring that the rights of others be taken into account is not interfering with innovation, it is a spur to safer innovation.

But instead of ethical production we have something that is the reverse of the precautionary principle. It's as if the real rule is:

---

9. TSCA requires companies to submit information they have on the health and safety of the new chemicals they want to introduce. In 2005, EPA fined DuPont the largest civil penalty until then, $10.25 million, for not reporting information they had on the risks of perfluorooctanoic acid, https://www.epa.gov/enforcement/reference-news-release-epa-settles-pfoa-case-against-dupont-largest-environmental. But as the movie *Dark Waters* shows, they continued to produce forever chemicals.

> When an activity raises risks, it should not be prevented unless the risks are certain and substantial, and will be borne by people who lack the political power to resist it.

We see the inability to really think about how we should regulate technology in the example of nanomaterials.

THE SAFE DEVELOPMENT OF NANOTECHNOLOGY

In the middle of the first decade of the third millennium (around 2005), I was asked if I wanted to participate in meetings convened by the head of research at the Massachusetts Department of Environmental Protection, Carol Rowan West, to look at "Emerging Contaminants"—endocrine disruptors, pharmaceuticals and personal care products (PPCPs),[10] and nanomaterials. I volunteered along with others from the state's Department of Health, the Department of Labor (workforce development), the Toxics Use Reduction Institute at UMass Lowell, and the Office of Business Development, to research nanomaterials. At the time, there was a lot of public fear about nanotechnology because of a Michael Crichton book titled *Prey* that imagined swarms of nanobots creating a "grey goo" menace. It is sad to think of how this fantastical scenario took over the public's mind, so that gauging the risk became an issue characterized by hysteria right away.

When we looked at what the scientific literature was showing, there *were* indications of concern. I particularly remember a picture of a carbon nanotube piercing cell material just like a spear of asbestos would. I remember a buckyball ingested by a microorganism that retained its stable spherical shape inside the thing.[11] Didn't seem all that digestible, but what could we conclude? More research is necessary, of course, but as far as we were concerned, there was urgency in making sure the industry developed safely, no matter what the risk might be. We looked at information about sunscreen that seemed to be emblematic of the risk we ought to attend to right now, without waiting: it seemed that nano-titanium in sunscreen could enter the bloodstream if you had a cut. We saw information that showed some nanoparticles could cross the alveolar barrier that was supposed to keep out everything but the gases you need for breathing and, once in the bloodstream, could travel anywhere in the body. There was a term for this: translocation. With nanotechnologies we might get increased translocation, as a side effect of using

10. See, for example, Christian Daughton's work: https://cfpub.epa.gov/si/si_public_record_report.cfm?Lab=NERL&dirEntryId=161023, or Rolf Halden's: https://biodesign.asu.edu/rolf-halden.

11. Also known as fullerenes, named after Buckminster Fuller, the developer of triangular forms in what are called geodesic structures, now engineered at the nanoscale out of carbon.

modern materials. Didn't we need to know this before we slathered this stuff on? Is it safe for my baby? Is it safe for the little fishes we love to greet in the kettle pond into which we place our bodies? (Materials that don't break down will accumulate in kettle ponds, which have no brooks carrying water away.) The other example that haunted us was people buying shirts and socks impregnated with nano-silver. This would kill the bacteria that causes odor. Then the nano-silver would wash out of the clothing and where does it go? I knew, from earlier work promoting the recovery of silver from photographic film wastes that, in its ionic form, it kills microorganisms. That's all right when it's your fork—but what about the discharge from washing your socks?

The world of governance, and those who watch and work with governance, can seem to break down into two categories: one group asks if we are seeing this stuff show up in our environment, and hearing that we are not (is anyone looking?) isn't sure there's a problem; the other group says what we are seeing indicates we will have problems if we don't act in some way.

Perhaps because of the fears Crichton raised, we watched a very interesting thing happen. At first, our searches for products containing nanomaterials found many bragging about it, including some that didn't seem to have any nanomaterials in them, but said they used nanotechnology in order to sound interesting and attractive. One product was a hair dryer! But that started going away as a result of the negative connotation that nano had acquired because of Crichton and soon we could hardly find anyone advertising that way, except for nano-silver, which persisted.

There was an extremely bright young engineer on staff at my office named Morgan Mihok who was able to learn about what was going on in the state from a consulting company that worked with companies using nano, and a company that made nanomaterials. Morgan found out that there was indeed a very active industry developing, but both of the companies were very protective of the identities of the companies doing this work. Wherever I was able to make inroads through my contacts with industry, I met with the same thing. With only a few exceptions, we found an enormous fear of public scrutiny. It was as if no one wanted anyone to know what was happening.

After the Office of Business Development completed a study that showed that there were indeed many companies working on nano in Massachusetts, their staff stopped coming to our meetings. It was as if the industry just wanted to pull a big blanket over themselves and operate in secret. Having worked for years on nuclear issues, I recognized this attitude.

About twenty years earlier, when I was writing *Radiation Events Monitor*, I had attended a conference on nuclear issues at which I struck up a conversation with a man from the industry. It was a friendly exchange and the man was in the middle of offering me a cup of coffee when one of his colleagues came over and said, "Don't you know that guy's from CARS (the Center for Atomic Radiation Studies)?" The offer of coffee, which I was reaching to accept, was withdrawn. I said, "But it's not an 'antinuclear' organization. It was founded by doctors who think the issue of radiation safety is important.[12] Don't you think that doing nuclear operations safely is in your interest?" The nuclear guy looked at me and said he knew that we weren't an antinuclear group. "But you're a close cousin!," he said. And thus he had to withdraw behind the defensive walls of the castle, in his imagined war between his enterprise and those who had no right to slow or stop it. The CARS discussion of preventing the risk as a shared concern, an offer to occupy a shared space in civic planning, was spurned by the industry.

It was a familiar feeling, for me, when I saw the fear of light that came with the nano issue. They didn't see that it was better to make sure things operate safely, and to let everyone know that they cared about that. They wanted to gamble on people being dull and stupid and not caring. This upset me, as the nuclear blindness did back then.[13]

This attitude was not universal. When I spoke to some attorneys, such as DC's Lynn Bergeson and at Boston's Prince Lobel firm, I found agreement about the benefits of looking at these issues, and was invited to mingle with businesses. But the avoidance impulse was powerful and widespread. I think of it as a felt sense of repugnance about the idea that government might even just turn its attention to them, so to counter that what was needed was the assertion of a context that would exclude government scrutiny. It was an attitude like a wall.

What can you do when you are a government that is friendly, and the target population doesn't want to be friendly with you? You have to come up with something that works for both sides—that allows you to assert and make manifest the public interest. In the meetings with my government partners, I suggested what we wanted was "Safe Development": to emphasize the idea that we didn't want to stop the industry but simply work to make it safe. We held two conferences using that phrase.[14] I spent an inordinate amount of time

12. I was actually personally opposed to any new nuclear plants and wanted old ones shut down, but this was the position of the center, as expressed by founder Dr. Donnell Boardman, and I thought it a useful one.

13. For an intimate portrait of nuclear workers, see Paul Loeb's 1982 *Nuclear Culture*: "If you try to confront the mission or how it's carried out, you risk not only being branded a pariah, but also being overwhelmed by the slow crush of bureaucratic resistance. So in this place (Hanford, Washington) where nuclear dangers are most intimately encountered, almost no discussion deals with their implications" (229).

14. "Getting the Big Picture: Safe Development of Nanotechnology" (2007), https://www.mass.gov/files/documents/2016/08/ud/safenano.pdf, and "Promoting the Safe Development of Nanotechnology" (2009), https://www.mass.gov/files/documents/2016/08/st/nanoproced.pdf.

on the phone trying to get people to come. That's another thing you can do when faced with an unfriendly group you need to reach: persist.

We got a couple of companies: Seth Coe-Sullivan of QD Vision and Mark Banash of Nanocomp, both proud of how they worked safely and completely contained the nanomaterials in their product without releases; Andrew Maynard, a world-renowned researcher on the topic; Matthew Hull, founder of Nanosafe, Inc.; and Diane Mundt of ENVIRON.

We got a few lawyers and people from insurance companies. But given how many were on the original list, the vast majority of the business itself stayed away, even though we featured a presentation by arguably the foremost authority on applied nanosafety, Chuck Geraci of the National Institute for Occupational Safety and Health, who explained that existing methods were applicable—that real safety was doable.

Being persistent with the Nanobusiness Alliance worked. VP Aatish Salvi, after many conversations, had expressed the fear that government involvement in the issue could ruin things, yes, but on the whole, this was a good thing we were doing. I had hit hard on the message: *we are not opposed to your technology; we just want it done safely*. In order for that to happen, we have to be talking about that in the open. We want the benefits of nanotechnology. We know it can give us great things. We know it can help with battery technology, that it can be used to target the delivery of medicines right to where they are needed. *We want those benefits, but we want them without unnecessary risks, and the only way that will happen is through the approach of safe development*. Government has no interest in shutting down this technology. We want it done right.

Salvi confessed to the crowd his initial reluctance to participate. But then he said he realized that our intentions were good. He said he agreed with our approach. He said he supported our effort. Unfortunately, within a year he was gone.

Vince Caprio of the Nanobusiness Alliance stayed in touch with me, and invited me to several events, and when I published the first (and only, to my knowledge) state guidance of its kind, entitled *Considerations for the Safe Development of Nanotechnology*[15] (with much help from Morgan and others), Vince shared it with his members.[16] Others praised the guidance, including attorneys working with companies doing nanotechnology, helping their clients do the right thing. The message may need to be repeated until it sinks in, but eventually, it can work. You put it out there and see if maybe it can catch on and spread. When it does, you are grateful to do the work the

15. https://www.mass.gov/files/ota_nanotech_guidance_final.pdf.
16. https://vincentcaprio.org/nanotechnology-considerations-for-safe-development-from-massachusetts.

government is paying you to do. Meanwhile I would not be telling you the whole story without noting the attorneys who argued for shifting focus away from the risks of nano. One stressed its similarity to natural nanoparticles and essentially argued that we were unfairly focusing on industrial sources. I told him I didn't see why similarity to another risk meant we shouldn't be concerned with it, especially if it is additive to natural risks we can't control.

OTA worked on nano for some time but got on-site only to a few facilities doing nano. As OTA's work is confidential, I can't say where I went, but it was a large facility and they had not yet started mass production, and were wondering what might go wrong if they did. I would love to name the company because they deserve praise for being anticipatory.

I looked at a large roller and asked where the coating with nano would be applied and imagined the particles flying off as they dried and asked questions like that. They nodded and took it all in and said they weren't doing anything yet, and wouldn't until they knew they could do it right. People should know that this is not common, but it is not that uncommon. There are companies like that, who do their best and recognize the value of the service government offers to help find the best options.

As mentioned above, I have a course at Boston University in which students do projects to help agencies do their work.[17] For years I would offer a project on nanotechnology, because I knew that work on that in the agencies had almost slowed to nothing. The strong support of former DEP commissioner Laurie Burt was no longer available, because she was no longer there; Morgan had left to become a doctor, Carol Rowan West retired, Dr. Michael Ellenbecker of the University of Massachusetts, a major source and supporter of this effort, retired, and as usual nano companies were not responding to OTA's offers of help. Finally, one of my students, Greylin Neilson, chose to work on nano, and she found information in the Boston Public Library's business reference section that showed that there were more than forty companies doing nanoproduction in the state. Sam Lipson of the Cambridge Health Department hosted a meeting of the few people left interested in this issue,

---

17. "Research for Environmental Agencies and Organizations." Projects are posted at www.bu.edu/rccp. The class grew out of a previous project called the Regulated Community Compliance Project, which compiled information about what states are doing to enforce the new lead poisoning prevention law, the Renovation Rule. It was easier to keep the "RCCP" label than to get permission to start a new website, and the lead poisoning prevention information is just as fresh and important as before, as what states are doing to enforce—or promote if they haven't been authorized to enforce—enforcement and education about the Renovation Rule remains terribly inadequate, as far as I can tell. It would be good for someone to repeat the research project we did to find out if that's truly the case, and to motivate people to pay more attention to this valuable tool for preventing the tragedy of lead poisoning, as our work seemed to do, a little bit, for what seems like just a moment. One state complained about what the data said (but it was submitted by them) and one seemed to step up its efforts as a result of the report.

including Laura Spark of Clean Water Action, which had been pressing for state action on the issue.

Clean Water Action has kept this issue alive.[18] It is through awareness that we avoid unnecessary risks. The precautionary principle means paying attention. It means being sensible. The idea that one should hide out and that the thing to fear is a fearful public is a recipe for disaster. It increases the risk that the public will freak out if they find out nano is happening near them, and it increases the risks to workers and consumers and the environment. The situation we have now of companies hiding out from the government and public scrutiny is not good.

I recently attended a Zoom workshop[19] with Andrew Maynard, one of the original investigators, and he talked about how he'd moved on from a focus on just nano and now put it in the context of many other risks. Guenter Oberdoerster,[20] who did some of the first experiments showing there are problems when cellular matter interacts with nanomaterials, noted the risks had been exaggerated. But both said we needed to know more. Both said that there *are* concerns. More to the point I have been making about the attitudinal wall I encountered, Vicki Colvin talked about how when she made presentations on nanosafety[21] and said that while nanotechnology can bring benefits, it also presents new risks, that people were upset by that.

Dave Rejeski, who created a list of products with nano in them while at the Woodrow Wilson Institute, had not worked on nano for years and had stopped updating the list. Some time ago, I wrote to him that the original website was hijacked by some company selling a health product. That was sad. But as I write this book and check on it, I see it is up and running again, and using crowdsourcing tools.[22] There has also been recent attention, workshops from the National Nanotechnology Coordinating Office and Harvard's

---

18. Full disclosure: Clean Water Action has funded my work on the issue. It is truly tiny funding. I can't afford to give lots of time away, but I can give some time away. A little grant, a small contract, can capture the motivated. Many people like me, retired from government service or just eager, responsible types, would work on small projects that can help address things agencies are not able to get to. A lot could be accomplished with small amounts of money for people like me who will gladly work for only a few dollars if it's something worthwhile to do.

19. NanoEHS webinar by National Nanotechnology Initiative, moderated by John Howard, director of NIOSH, https://www.youtube.com/watch?v=ZcdQ34k5oVg.

20. "We investigate the propensity of these particles of different shapes (e.g., spheres, tubes, rods), different chemistries (e.g., metals, semiconductors, carbon) and different surface characteristics (coating, charge, porosity) to translocate from the site of deposition in the respiratory tract to extrapulmonary organs such as heart, liver, bone marrow and brain is being studied," https://www.urmc.rochester.edu/people/20180430-guenter-oberdoerster.

21. Then of Rice University (now at Brown). See, for example, this 2006 presentation at an EPA conference, which contrasts a picture of Tinkerbelle, representing the good that nano can do, and Darth Vader (23): https://archive.epa.gov/oswer/nanotechnology/web/pdf/id-introduction-presentation-slides.pdf.

22. http://www.nanotechproject.tech/cpi/browse/.

TH Chan School of Public Health.[23] While the risks were not as immediate as they might have seemed to some, they remain an important area of focus for many.[24] In 2021, Clean Water Action and I put on a series of Zoom calls for health, fire, and other local officials. Attendees reported they were glad they attended, having known nothing about the topic beforehand and agreeing it is important to know about. We are providing the basics of nanosafety, but it ought to be the government that does the work we are doing to fill a gap. Clean Water Action's tiny grant to work on this issue has run out.

Back in 2011, I followed up the two conferences introducing the concept of safe development with an *Invitation to all Nanomanufacturers or Researchers*:

> Be A Founder of the Manufacturer's (and Researchers) Environmental, Health and Safety and nano-enabled Green Alternatives Forum (the MEGA Nano Forum).

The MEGA Nano Forum was coming together when I was required by the state to disinvest in it and pay attention to other matters, as a response to budget cuts and reductions in staff.

When government disinvests, it loses a great deal. It could find students, retirees, volunteers, contractors, nonprofits to continue the work, remaining as the fostering center. But too often, new leaders ignore what has already been invested and move on. When you have short resources, you choose only the highest priority items and the ones not rising to that level never get addressed. We didn't have enough staff to do all the work.

But the idea was well received and distributed by lawyers, government officials, and consultants to the industry. It got support from people in the insurance world. I had a LISTSERV, Safe, and Green Nano, and had a couple dozen on it for a while. I was invited to conferences to discuss these ideas:

- Paying close attention to reducing risks from nanoparticle exposures and releases is critical to maintaining acceptance of the technology.
- Uncontrolled risks will hurt development.
- A pro-safety, pro-environmental, open process is also pro-development.

---

23. See, for example, "Consumer Exposures to Laser Printer–Emitted Engineered Nanoparticles: A Case Study of Life-Cycle Implications from Nano-Enabled Products," Pirela et al., https://pubmed.ncbi.nlm.nih.gov/25387251/.

24. At the April 27, 2021 National Nanotechnology Initiative webinar, "What We Know about NanoEHS: Environment," some panelists emphasized that risks have not, generally shown up at a more than "low" level. The Harvard program has produced information on new risks, such as Sandra Pirela's findings of nanoparticles released from laser printers, https://www.hsph.harvard.edu/news/features/laser-printing-has-gone-nano-but-is-it-safe/.

At these conferences, I spoke frankly to the nano-people about how there was a widespread attitude in the nano-world against engagement with government and with openness but that this was not optimal. We need to know and then if necessary to regulate and when we do it has to be done right. Not helping people to know what's going on—what nano is in, for example—ensured ignorance. Not engaging with governance was a way of ensuring it could not be done right. I didn't quite mock "Fear of the word Precaution" but came close. I said an assumption that any regulation would be bad was not borne out by history.

At these conferences packed with technical presentations, I was very often the only one to explicitly focus on the moral issues involved.[25] I talked about the bases that need covering. These are the potentially affected populations—"stakeholders"—we need to be thinking about. I would throw in: because we should care about them. But I would also base it in cold commercial logic: if you harm them, they might sue you; they might raise a panic about your operations. I would discuss the issues in moral terms but ground my argument in the business logic of a preference for public acceptance, avoiding interruptions, and avoiding liabilities. I tried to conjure up the picture of the flow of money stopping once your technology is perceived as scary and the best way of addressing this business interruption would be an open and verified commitment to safety.

One question was what to disclose, if anything, about contents or operations, and to whom. It was pretty clear that just keeping everything secret was not the answer. I would talk about the "Rule of Empathy": if you are a maker or proponent of nanoparticles, imagine that you are a consumer—What would you want to know about what you are buying? That would be a pretty good guide to what you should disclose. I would note that providing appropriate warning according to this rule reduces potential legal liabilities.

I often showed this graphic, shown in figure 6.1.

The point about disclosure is a theme I have used for many years in training I give to real estate professionals, and it is backed by countless advisors to the real estate professionals: disclosure shifts responsibility. If you don't tell a buyer about lead paint, or a broken furnace, or a leaking roof, they may sue you if there is subsequent harm or damage. If you do tell them, it's on them that they didn't do something about it. You are home free.

It felt like all of this was received extremely well. My takeaway was a good one: that people like hearing basic moral principles. But I found myself increasingly depressed by the experiences, as what it said to me was that these simple ideas needed to be said—and had not been.

---

25. There were some striking exceptions. Barbara Karn of EPA brought a strong moral perspective to the creation of the Sustainable Nanotechnology Organization, http://www.susnano.org/governance-karn.html.

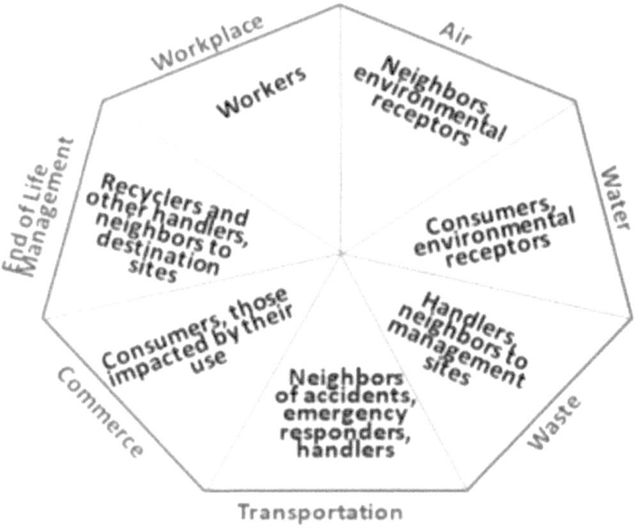

Figure 6.1. The Seven-Sided Universe and Stakeholders in Each Dimension

It was like the expert money hunters who ran companies and had never realized how much they could save over the long run by spending a little more up front to avoid the use of toxics. They had never hunted in this part of the forest before. While everybody was off calculating numbers about risk, there were some really simple things to do that you should just do without needing any numbers, like using filters on the vents and managing them safely after use.

In all the public discussion of nanotechnology, it seemed to me that the focus was on the technical characteristics. In presentation after presentation that was focused on safety, I heard highly technical approaches based on highly specialized analyses. These were all quite important, as the details are critical. But the overarching framework was only dimly perceived.

I tried to make only simple points, in contrast, as if to balance things out. I would say that all the complex differentiations between types of nanomaterials, with which most presentations began, or the decision-matrices that one would use to deal with them, or the risk assessments on which those decision-matrices were based, could be structured around a very simple framework of three types of nanomaterials:

1. Free nanoparticles—"instantly releasable"
2. Released during use or under certain conditions
3. Safely embedded until end of life

In contrast, I often find the discussion framed by incredible complexity. Experts can handle those complex aspects but we need widespread clarity, and the complexity can be reduced to a manageable level, as the above formulation, which guides appropriate response to the risk.

Anticipatory government should facilitate the needed conversation. We could be developing clearly applicable policy that would be widely understood and applied. That happens when the perspective is general, shared, open, and universal—not just the specialized.

The limited focus on making product that will sell, excluding notice of the life-cycle impacts or potential harms of those products, means that we don't know what nano is in. The worker who has to deal with it may be getting a safety data sheet that has the same information as the larger version of the substance. But when you engineer materials at the nano level (a billionth of a meter), characteristics of the material are altered. Reactivity increases, as there is more surface area to react with. Gold can be red. Things that the body kept out, might now get in. Things that don't normally catch fire, or explode, might.[26]

We developed training for local officials because the federal and state governments have failed to act sufficiently. In our training we note:

> Until the national government really kicks into action,[27] states and municipalities should take action. Municipalities should know what questions to ask, if a nano-using or producing company applies to locate in town. Their permits should address the risks. The planning board should make sure it is informed before approvals are made. The health board must know how to answer questions from neighbors, and if there is no environmental officer, to make sure there are not releases. If the municipality has no role in protecting workers, it should ask the state or federal agency to step in to check things out. The municipality must know how to respond if there is a spill, fire, or explosion. Local medical facilities need to know how to respond to injuries. These may all be unlikely events. Much of the industry I hear about and certainly all that I have seen, is doing things safely. *But the history of industrial accidents and pollution is one of wrongly assuming that because it's mostly safe, it's always safe. Attention must be paid.*
>
> If the feds and the state aren't doing it, municipalities should require facilities to report essential data about materials, processes, and safety protocols. They should find out what nanomaterials are being used or manufactured on site,

26. https://www.nfpa.org/-/media/Files/News-and-Research/Fire-statistics-and-reports/Hazardous-materials/RFFireExplosionHazardNanoParticles.ashx.

27. EPA has issued some rulings, preventing some risky practices, primarily through the Toxic Substances Control Act, but nothing comprehensively preventing risks; and has issued a call for information that was either largely ignored, or the results have been kept under wraps.

> and if they have combustible or reactive properties. They should ask questions: what are potential hazards that could cause harm to workers and how does the facility address those potential hazards? How might materials be released—and what has the facility done to prevent/control release? What does the municipality need to know about materials to respond effectively in case of accidental release, fire, or explosion? If an inspector walks in the door, will they be provided with the proper protective equipment?
>
> How does the facility prevent releases or exposures of nanomaterials to air, during all phases of operation, from the arrival of the raw materials at the facility to the containment of materials for transfer from the facility? If processed under containment or negative pressure ventilation, how do they monitor to ensure proper functioning of the system to prevent all fugitive releases? Do the ventilation system filters prevent all releases to the environment? Does the facility use methods other than wet washdown, to prevent nano in wastewater discharges?
>
> Municipalities can and should ask the facilities to describe how they prevent releases to water or soil. How is waste handled? Are waste handlers warned, are the particles sealed somehow till final, contained disposition, so it will not be dispersive?

Nanotechnology may give us better solar power, better electrical cars, and better medicine. *We want what nanotechnology can give us. We need what nanotechnology can give us.* But we don't need paint that has nanomaterials if they will be coming off the walls. Even though buckyballs may increase the lubricity of motor oils, we don't want applications that will disperse, as buckyballs would from cars using it in their motor oil. Even though carbon nanotubes may allow us to make superstrong materials, or to conduct electricity without using metal, we don't want workers breathing carbon nanotubes. The analysis is complex, but the question, the goal, the point, is simple.

We can have the good without the bad, but only if we take actions we have still not taken even though there has been public discussion of these issues for years. We should not expect municipalities to be able to take this action. We need a concerted approach from all levels of government.

We live our lives in clouds of dusts, and they are not the dusts of old. They are the dusts of manufactured items made with new technologies. We live among, breathe, and ingest fibers and all manner of plastics, and the bits of tire that wear away from our cars, and weathered particles of paint, and the chemicals spread on the lawn, the pesticides in our food, and the preserving solutions sprayed on our fabrics, with particles large and small. There are things we can do about all of this. Paying attention to nanotechnology should not make you freak out if a nano-company comes to town. They may even be a model for others to be like, and they may produce something fantastic while providing jobs and boosting tax revenue. But that benefit is not an excuse to ignore or accept risks that need not be posed.

At an April 27 webinar by the National Nanotechnology Initiative, titled "What We Know about NanoEHS: Environment," one speaker noted very low risks showing up in crops, but there was something about copper, if weathered, taking on a different valence and reacting with sulfur. There was much in terms of gaps in our knowledge, but thus far no alarm bells and many potentially positive applications. We're in go mode. Maybe it will go well. But we should have in place required reporting in recognition of the right to know of communities and individuals about new risks that may be imposed upon them, and how those risks are managed. We need guardrails on those high mountain roads.

OTHER EMERGING CONTAMINANTS

Genetically modified organisms represent a similar problem. There is much that genetic modification can do that we want. Good GMO is clearly seen in the response to the pandemic. Had we not learned how to make vaccines that use messenger RNA, we would not have the vaccines that now give us all some hope for a return to normal life. But then there is GMO that seems to be for the purpose of gaining control of a market. When you are selling glyphosate, and also genetically modified plants that are resistant to glyphosate, and you have obtained patents on those resistant seeds, and farms killing weeds and using resistant plants have higher production than those not, this resembles a plan for capturing the food production processes of the world. Another reason for using glyphosate is to simply dry out your crops quickly, so you can get them more quickly to market. That's good for a mass production industry but not the long-term interest of farmers, food consumers, the soil itself, the water into which the chemical runs off, the many species of life that are harmed by it, or the ecosystem in which food production occurs.

GMO technology is another example, of many, of capabilities resulting from substantial government support for basic research. Those are intended to be shared with the public at large, but whether this produces widespread, lasting benefit to the public or are captured for private benefit by a limited set of parties is a legitimate question for public consideration.

Unfortunately, we tend to think and talk about these issues in terms of generalities and to quickly take sides and dispose of the question. The industry points to good GMO to say that genetic tinkering is good, and others point to bad GMO to say that genetic experimentation must be stopped. It's not a binary good/bad decision but a complicated question involving much uncertainty. It's not a discussion to have once and be done with, having taken sides. It ought to be a continuous effort to develop a sophisticated understanding of what is permissible or good technological application and what crosses a line, and design our regulatory responses accordingly. We need to keep thinking about that line, and all its gradations, and all the varieties of response of encouragement and discouragement.

In the case of genetic tinkering an ongoing effort is critical, because we are messing with the mechanics of life itself. A kind of consensus may be developing that human reproductive DNA should not be altered, for that may affect future generations, while human DNA can be manipulated to help someone already alive from continuing to suffer from a curable disease. Where the line should be drawn concerning what's needed and desirable concerning agriculture is not as clear, with "No GMO" possibly fighting a losing battle as pollen from GMO crops frequently contaminate their loads and if the costs of raising food that will pass as without GMO increase. The discussion calls out for sophistication.

That courts have allowed the patenting of life is a question that merits public discussion. Inventions are patentable, but natural phenomena are not.[28] But when our powers of inventiveness begin to alter the mechanisms of life, our focus must be expanded to see the impacts on the whole. The Court has examined the question under patent law of whether some new useful improvement has been made. It has not legislated with an eye to the future but applied existing law. But a larger view, that the public should take, would consider impacts when newly created life-forms change ecologies. It is also not unreasonable to consider the product's usefulness in an expanded context, who would use it and for what purpose. Could it bring food production under the control of just a few, producing misery among small farmers all around the world? The idea of patents is to promote innovation, to advance science. Patent law is a tool of governance that can be used to promote or stifle innovation. Entities that have gained power over food production systems may be able to generate sufficient revenue to fight competition, preserve their hold, and prevent innovation.

---

28. In *Diamond v. Chakrabarty*, 447 US 303 (1980), the court allowed the patenting of a life-form that was not found in nature. In *Association for Molecular Pathology v. Myriad Genetics* WL 2631062 (2013), the court said isolated gene sequences were not patentable, but altered ones are.

ENDOCRINE DISRUPTORS

Chemicals that disrupt our hormones—by mimicking them, inhibiting them, or changing how they function in the body—are analogous to delinquents who point lasers at aircraft, hoping to blind or confuse the pilot. These chemicals disrupt the signaling of our body's chemicals, and cause damage we have trouble predicting. Damage from exposure may not exhibit the same dose-response curve as ordinary toxins do. You may be at greater risk from a smaller amount than a larger. The time that you are exposed is crucial. If the disruption is of the development of a part of the body during the fetal stage, if you are hit when that body part is being formed, you may have a deformity. If you are hit after it has already been formed, you may not see an impact.[29]

It is important to consider that many endocrine disruptors are chemicals we probably don't need. They leach out of plastics, and were put in so the plastic would have better characteristics, such as pliability, softness. Convenience and quality of product is not the only characteristic we should care about, because that they will leach out and will cause mischief to living systems can be foreseen. They leach out of sofas, where they were put in for stain resistance. They leach out of many things in which they were placed as flame retardants, but studies have shown that they hardly increase safety.[30] Though we lack the science to really know the damage, we should be able to grasp the clear moral principle concerning the imposition of these risks. It argues that endocrine-disrupting chemicals should be banned unless they are used in ways that prevent their release, so that humans and other life-forms are not subjected to them.

OTA received a request for help from a company using BPA, when the BPA scare first hit, and companies were selling "BPA-free" product. We gave them a bunch of information and some recommendations, and then they let us know they were going with BPS. They switched from bisphenol A to bisphenol S so they could claim they were BPA-free. BPS was a close analogue. We wrote them a letter advising that this might make them vulnerable to criticism, and encouraging them to continue working with us on real alternatives, as the risks of BPS could likely turn out to be similar to the similar chemical BPA. It was one of those letters we had to craft carefully as we want

29. See the work of Theo Colborn, https://endocrinedisruption.org/about-tedx/theo-colborn-ph.d.-president/, and the 1996 *Our Stolen Future*, written with Pete Myers and Dianne Dumanoski.

30. See, for example, the series begun in 2012 by the *Chicago Tribune*, "Playing with Fire," http://media.apps.chicagotribune.com/flames/index.html, and the work of Arlene Blum, founder of the Green Science Policy Institute, https://greensciencepolicy.org/about/.

to keep working with companies, but we have to tell them things straight. We didn't hear back. BPS is now a poster child for "regrettable substitutions."[31]

Some parts of industry have benefited from a regulatory approach that demands evidence that substantial harm is occurring before restrictions come into play. The Toxic Substances Control Act allows EPA to order a company to perform tests, but EPA cannot do that unless there is evidence it is warranted. How do we get the evidence? The strategy evinces an unmerited wishful thinking that it will simply appear, but when most toxicologists are not finding work in government that has reduced funding for such work, and must work for companies, this is not as likely as it would be if there was widespread funding for independent testing. Those who wish to see governance reconstructed sensibly should call for vast increases in funding for independent science and permit approvals based on pre-market evaluations.

The precautionary principle's application to this category of risk is clear. Given the fact that we may not recognize the damage, or that these chemicals have caused visible damage, logic tells us these chemicals will impose a certain amount of damage, and the best science we have confirms this logic. We should act on that understanding. The old idea that business priorities must come first, above the integrity of our bodies' functioning, must be discarded. Our health and the functioning of all endocrine systems—including nonhuman—must be the priority. Business needs can be accommodated within that prime directive. This is a simple point that could be better understood. We don't need to be experts in biological science to see it and asserting it does not make anyone antibusiness.

FOREVER CHEMICALS

Some endocrine disruptors are also "forever chemicals," a new term that joins the old, Persistent, Bioaccumulating Toxins, (PBTs). We began the practice of regulating PBTs after learning how PCBs—polychlorinated biphenyls—did not break down. We also refer to "POPs," as in the International Treaty on Persistent Organic Pollutants.[32] We have the concept of chemicals that don't break down, but we haven't employed it to prevent such chemicals from being produced and distributed.

---

31. Leonardo Trasande, "Exploring Regrettable Substitution: Replacements for Bisphenol A," *The Lancet* 1 (June 2017): e88–89, https://www.thelancet.com/pdfs/journals/lanplh/PIIS2542-5196(17)30046-3.pdf.

32. http://www.pops.int/Home/tabid/2121/Default.aspx; https://www.epa.gov/international-cooperation/persistent-organic-pollutants-global-issue-global-response.

In my first year working at the state as an environmental analyst I told my superiors that I saw evidence that electrical transformers had leaked PCBs,[33] and that the law said that has to be addressed immediately. This is because it had become known as a chemical toxic to life because it doesn't break down. Strict regulations and bans on most uses were enacted under TSCA (one of the few times the law's powers have been utilized, and one of few times we have been able to see how effective bans can be compared to halfway compromise measures). But we did not learn our lesson. *After the PCB ban, we should have made sure that no similar kinds of chemicals were produced.* We had the power under TSCA to prevent them from reaching the marketplace. As noted above, we failed to extend the lesson from PCBs to PFAS and they have found many uses, some dispersive and some trivial. Now they are found in our bodies.[34] Was it not clear that they were persistent?[35] This is an important question to consider: how to redesign the regulatory review so that it will catch such things.

The most important thing to understand about these chemicals, now costing us untold millions in treating water, was that this was the result of our failure to regulate, when we at least theoretically had the authority and we should have had the justification. We must make the attitudinal and philosophical changes that will allow us to ban things that should not be in commerce. We know—it is a simple concept—that living things need to metabolize what enters their systems, or excrete it safely—and that chemicals that do not break down must not be released into the environment so that they will be taken up by living things. The very existence of forever chemicals is dramatic evidence that something more powerful than common sense has captured our system of laws. Rebuilding our system of governance requires putting people and other living beings first.

---

33. Specifically, in electrical equipment at a state prison under guard towers, in spaces that flooded, dispersing the chemical.

34. "CDC scientists found four PFAS (PFOS, PFOA, PFHxS or perfluorohexane sulfonic acid, and PFNA or perfluorononanoic acid) in the serum of nearly all of the people tested, indicating widespread exposure to these PFAS in the US population," https://www.cdc.gov/biomonitoring/PFAS_FactSheet.html#:~:text=By%20measuring%20PFAS%20in%20serum,%20scientists%20can%20estimate,exposure%20to%20these%20PFAS%20in%20the%20U.S.%20population.

35. EPA did take some action. EPA fined DuPont the largest civil penalty by 2005, $10.25 million, for not reporting information they had on the risks of PFOA, as required by TSCA: https://www.epa.gov/enforcement/reference-news-release-epa-settles-pfoa-case-against-dupont-largest-environmental. For all appearances this was not enough to induce the responsible behavior we need. A reading of TSCA indicates that EPA could have done more under the law, but was that really true, given the political context at the time (and still, today)? It is not clear that an attempt to ban categories of chemicals can yet succeed, though the facts show it is a necessity if we are to protect the biology of living things.

NONIONIZING ELECTROMAGNETIC RADIATION

You have no doubt seen the commercials about what a wonderful world we will have because 5G is coming. What will it let us do? We can download a game or movie from anywhere, walking down the street! How wonderful. But is it truly safe to bathe our entire environment in strong electromagnetic fields? Those who say that it's not are often derided as conspiracy nuts, hysterics, and told that there is no scientific evidence for concern. Cell phones have not caused an epidemic of brain tumors, they say, and no one should stand in the way of the advance of technology.

But the fact is that hundreds of studies do show cause for concern.[36] And some applications of 5G are trivial. Yes, there are some good things that can come from it, but there are some bad, such as universal surveillance, in addition to the effects on our bodies. There is simply no reason to consider nonionizing radiation to be innocuous. After all, our bodies do have electromagnetic fields. While 5G is a millimeter wave frequency and likely does not penetrate to the brain as cell phone radiation may, there are questions about its effects on the skin,[37] and there is no reason to think that powerful artificial fields are completely safe.[38]

What seems to be the real force driving the rabid pace at which 5G is being pushed? Americans for Responsible Technology state:

> The primary purpose of 5G is not for better phone service, or for tele-medicine, driverless cars, the Internet of Things or emergency services. It's to allow wireless companies to compete directly with cable companies to stream video into your home. Wireless technology is not a utility.
>
> American telephone customers have already paid billions for a national, fiber-optic network that would provide reliable, low-cost, secure and safe broadband internet access to every American. But instead of using the funds to build the

---

36. For example, tumors in rats from cell-phone radiation: https://www.niehs.nih.gov/health/materials/cell_phone_radiofrequency_radiation_studies_508.pdf.

37. "Since the NTP's (National Toxicology Program) studies have demonstrated that there is some interaction between RFR (radiofrequency radiation) exposure at the tested frequencies and cancers of certain tissues, there is a need to understand the interaction between RFR and biological tissues and the factors that affect that interaction" (ibid.).

38. For a comprehensive summary of the issues see Barbara Koeppel's "Wireless Hazards," *The Washington Spectator*, December 28, 2020, https://washingtonspectator.org/wireless-hazards/:

> Warnings from industry executives such as Frank Clegg (Microsoft Canada's former CEO) are rare. So, too, are those from governments, since the industry lavishes huge sums on the lawmakers. According to the Center for Responsive Politics, from 1989 to 2017, the industry gave $101 million to members of Congress and their PACs. Moreover, the National Institute on Money in Politics says industry lobbying groups plowed $93.7 million into local elections in 2018.

network as they had promised, the telecoms diverted the money to build out their wireless networks, which are more profitable.[39]

In addition to this, many working on this issue point out that due to permissive FCC regulations and the 1996 Telecommunications Act that was intended to make sure everyone had access to the internet, companies can exercise a right to place wireless equipment wherever they want, unless a community promulgates an ordinance that directs placement under the constraints of the law. In my town, attempts to direct the attention of the select board to this issue compete with many other priorities, and not every member of the board, or the relevant offices or committees, has the knowledge necessary to know that swift action is required, before wireless companies come to town.

Some of the companies moving to install wireless equipment are small, and hope to rent space to the carriers of phone and other services later. The build-out is not happening as a result of careful planning, where equipment that might be noisy, dangerous, unsightly or harmful to your health is only placed where it's needed. I was told by the one select person who is responding thoughtfully to this issue that he was told by others on the board that this was an issue for those people who wear tinfoil hats. To them it's a crackpot concern.

Here we have another example of commercial interests—in this case large and small, organized and wild—weighed more heavily than that of health, or the reduced property values that homes near wireless equipment now experience. We have the presumption that raised concerns are better ignored, the same phenomenon of fearing precaution. Again, we have a willing blindness that prevents reasoned deliberation and planning by the community in default to planning by commercial interests. How many communities have staff or volunteers who have the time to become informed about this, and know how to take action? We need to amend the Telecommunications Act to put in place a process fairer to communities, using updated science. We need to rebuild governance without fearing the requirement of safety first, as it will not just reduce harm but help ensure good planning.

WHAT NEEDS TO CHANGE?

The issues discussed above are examples of what happens when a society values economic growth over other concerns. It is certainly true that technical innovations lead to commercial opportunities, and the age-old greed

39. https://www.5gcrisis.com/5g-basics. This association of organizations and individuals posts scientific studies that show cause for concern at https://www.americansforresponsibletech.org/scientific-studies.

factor comes into play. But there is something deeper going on, a love of technological advance, and a faith in it—an addiction to it. This is closely tied to an overweight respect for expertise, and a system of meritocracy. Louis Mumford is one modern thinker who has written about this basic orientation (or, misorientation), for example, in *The Myth of the Machine: The Pentagon of Power*, which depicts our modern worship of technology, and the power it brings, as strikingly similar to ancient sungod and divine king worship, and as a kind of addiction. Mumford recalls Goethe's fable of the Sorcerer's Apprentice—the idea that automation will free us from labor, but without the master magician's foresight and ability to correct out-of-control mechanisms, we are, like the apprentice, drowning.

> And as our knowledge of isolatable segments and fragments becomes infinitely refined and microscopic, our ability to interrelate the parts and to bring them to a focus in rational activities continues to disappear.[40]

While examining Leonardo da Vinci's life, Mumford says that he was different from the modern technologists who look to him as a model of invention. Mumford notes that he thought widely, not narrowly, and exercised caution and restraint, and that had we proceeded as he did, "naturalization, mechanization, organization and humanization might have proceeded together . . . had his moral sense not been awake, he would not have suppressed his invention of the submarine, because he felt that the soul of man was too devilish to be trusted with it."[41] Mumford states:

> What one must give up, in any effort to achieve a many-dimensioned and coherent world picture, is the idea of early achievement and instant exploitation. Whatever the field of invention, or organization, one must be ready to go forward at a slower pace, looking before and after; to make fewer discoveries, to spend as much time assimilating knowledge as in acquiring it; to do less, perhaps, in a whole lifetime in any one department than the concentrated specialist is able to do in a decade. From the standpoint of the power system this demands an impossible sacrifice: the sacrifice of power to life.[42]

The agenda of those who are willing to make the sacrifice that Mumford points to, who are not so addicted to the pursuit of power or even more modern luxuries that they find sacrifice of the pace of progress impossible to make, must include the design of legislation and regulation so that the precautionary principle is given meaning. In order to garner the political will to do this we may address the fears of those who irrationally fear government

40. Originally published in 1964 (1970 edition), 180.
41. Ibid., 160–61
42. Ibid., 163.

action by making clear that our new versions of the rules may impose a bit of slowing down but will not prevent technical innovation, will not prevent commercial development, and enhance the chances of beneficial progress.

Of course, governance can go wrong, but it can also go right, and we really have no choice but to take that risk, equipping ourselves as best we can against the chances of failure, with the vision of collective wisdom before us[43] and the respectful processes that increase our chances of manifesting it.

QUESTION

Do you know what's in what you buy? Are there nanoparticles in there, or forever chemicals, or endocrine disruptors, or GMOs? How would you know? What systems should we have in place to let us know? What do you say to someone who says we ought not to have this information?

43. Perhaps the most important US articulation of that vision is found in the work of John Dewey. See, for example, Tristian Stobie's appreciation of his book *Democracy and Education*: "Democracy is not only about extending voting rights, a big issue in 1916, but also equipping citizens with the ability to take on the responsibility to make informed, intelligent choices and decisions leading to the public good. He believed that democracy is not just a political system but an ethical ideal with active informed participation by citizens," https://blog.cambridgeinternational.org/reflections-on-the-100th-year-anniversary-of-john-deweys-democracy-and-education/.

Chapter Seven

The Quality of Our Awareness

Envisioning Future Programs, Remembering Forgotten Ones

REORGANIZATION

J. Clarence (Terry) Davies coauthored the plan for creating the Environmental Protection Agency. In 2009, the Woodrow Wilson Institute published his new vision for federal environmental protection.[1] It begins:

> Since 1980, the capability of the federal agencies responsible for environmental health and safety has steadily eroded. The agencies cannot perform their basic functions now, and they are completely unable to cope with the new challenges they face in the twenty-first century.

His proposal to combine environmental and consumer protection and base them both on a more robust program of scientific investigation was hailed in the preface as an important contribution by the first EPA administrator, William Ruckelshaus, who wrote that "We are smart enough to take advantage of the fruits of technological advances and to minimize or eliminate risks to people and the environment. But we need to learn from past mistakes and be able to anticipate future challenges." It is reasonable, based on modern achievements, to believe that we are smart enough to do this, but the question of whether we are smart enough to resolve to do so remains in question.

Davies's report uses the example of nanotechnologies, which manipulate matter at the molecular level, to point out that oversight capable of controlling risks is not just essential for public safety but also "to promote the development of the technology."

---

1. J. Clarence Davis, *Oversight of Next Generation Technology, Project on Emerging Nanotechnologies*, Woodrow Wilson Institute, April 2009.

> The "cowboy ideology" that views regulation as antithetical to free markets has proven to be false in sector after sector. Productive markets require effective regulation.[2]

Davies envisions an agency that would be funded adequately to be able to do the science necessary to evaluate risks from products, pollution and in the workplace, to monitor and do the research that can enable us to anticipate what's coming as well as what's happening, and to perform oversight in a much more integrated and synergistic manner than we do now. He would bring together the EPA, the US Geological Survey (USGS), the National Oceanic and Atmospheric Administration (NOAA), the Occupational Safety and Health Administration (OSHA), the National Institute for Occupational Safety and Health (NIOSH) and the Consumer Product Safety Commission, and call it the "Hypothetical Department of Environmental and Consumer Protection." This would enable the combination of the departments in each of those that look at risk, and the health monitoring necessary to understand risk, so that it can be seen over its life cycle and in all the ways it presents. It would also enable the combination of compliance, also most sensibly done looking at the entire picture at once.[3]

Such a large agency, Davies writes, would have to ensure the independence of each division, as well as their integration. With a better understanding of risks, certain products could be regarded as needing no or little oversight, and promising technologies that pose little risk could receive support. Right now, we have a piecemeal approach that leaves us without "any coherent strategy for the overall technological future of the world."[4]

In 2018, the EPA Alumni Association formed discussion groups to envision "The EPA of the Future." Referring to National Academy of Sciences reports on the need to boost scientific capacity,[5] the discussants envisioned a future in which

> Environmental right-to-know will have advanced to become the expected norm. All members of the public will have ready access to useful, understandable information about what is in their air, water, food, and products on a real-time basis and with reliable forecasts as well. Ubiquitous sensing will advance the ability to know all of the sources of contaminants and stressors and enable informed efforts to reduce and eliminate environmental hazards. Transparency

2. Ibid., 21.
3. Ibid., 26.
4. Ibid., 31.
5. *Science for Environmental Protection: The Road Ahead*, National Research Council (Washington, DC: National Academies Press, 2012); *Our Common Journey: A Transition Toward Sustainability*, National Research Council (NRC) (Washington, DC: National Academies Press, 1999); *Preparing for Future Products of Biotechnology*, National Research Council (NRC) (Washington, DC: National Academies Press, 2017).

> and clear communications will build understanding and confidence and a sense of teamwork rather than adversarial conflict. EPA will be recognized as a provider of environmental solutions, with the imposition of regulatory obligations and enforcement actions the failsafe mechanism that assures protection when markets fail to respond to open information.[6]

In order to achieve that future, it will be necessary to understand not just climate change but also the many stressors we experience. The needed scientific work includes grasping the complexity of "the chemical, biological, and physical characteristics of an agent, the genetic and behavioral attributes of a host, and the physical and social characteristics of the environment." We have tens of thousands of chemicals that currently can be produced for commercial purposes, and added to the complexity of understanding the impacts of each, they also present themselves in mixtures. We have an aging waste water and water supply infrastructure. We have not solved the problem of lead. We don't yet understand the "risks of trace pollutants originating from prescription medicines and other biologically active substances." Don't forget air pollution, nutrient pollution, disturbances of the land/atmosphere interface, or the impacts of the agricultural sector on water quality. The job of controlling industrial discharges is not finished, nor the job of understanding ecologic impacts. The group noted "Continuing and increasing leakage into the environment from waste disposal sites that were believed to have been adequately sealed and capped, along with new types of disposal challenges posed by the dramatic expansion within the plastics industry and the increased extraction, use, and disposal of rare earths and other previously unrecognized hazardous materials."[7]

Ex-EPA staffers who care enough about the problems they worked on in their careers to continue to provide their expertise should be listened to. There's a lot to do. Ruckelshaus, their leader twice, says that we have the scientific capability to do it, and/or the capability to develop the science we need done. The EPA Alumni sketch out a feasible program of protecting the atmosphere and our waters, protecting ecosystems and biodiversity, improving urban infrastructure, and protecting citizens,[8] but where is the political will? The alumni advise that we should have a goal of national well-being, rather than growth in the gross domestic product (GDP), and that EPA must improve its communication with the public.

---

6. Focus Group 4: Science, Technology, and Information, https://www.epaalumni.org/userdata/pdf/FG4.pdf, 2.
7. Ibid., 3.
8. Focus Group 1: Future Environmental Challenges. https://www.epaalumni.org/userdata/pdf/FG1.pdf.

Meanwhile, support for environmental protection has not disappeared. In general, though it varies from state to state, a majority of Americans think environmental protection is worth the cost.[9] But Americans have allowed their elected leaders to bring about the steady diminishment of our environmental commitment. In 2018, senior communications director Keith Gaby of the Environmental Defense Fund wrote:

> In 1979, the agency's budget was $5.4 billion. In 2017, its funding was $8.2 billion. But adjust that for inflation—so you can actually compare what EPA had to spend on enforcement, state and local support, science and all its other responsibilities—and the 2017 budget was really worth just $2.4 billion in 1979 dollars.
>
> This dramatic drop occurred even as our population increased by one hundred million people and our economy doubled in size. That means that with many more people and businesses, the agency has far fewer resources to protect clean air and water, deal with toxic waste and chemicals, and protect our health.[10]

Plus, EPA has had many new responsibilities to deal with. While the discussion about EPA's future rightly includes reorganization such as Davies suggests, and improving the scientific and communications capacities as so many other experts say, it's necessary to see the shrinking of resources and how that has limited EPA's ability to communicate with the public, and how that has contributed to a diminishment of support. The quality of our awareness should encompass the ability to understand why our laws don't seem to be working, as well as what our future system of law can be.

NOT WORKING AS INTENDED: THE LEAD DISCLOSURE RULE

The experience of providing training for real estate professionals on environmental issues in residential real estate, primarily on lead, showed me a very low level of awareness—of the problem, of the requirements, of their meaning, and of the consequences of not understanding the law or its purpose.

When I enforced the rule that lessors and sellers must disclose information about lead in the residential properties built before 1978 (for EPA, from 2001 to 2003), I repeatedly encountered the idea that the violations we found were just mistakes in paperwork, when they were actually about increasing the risk

9. Pew, https://www.pewresearch.org/fact-tank/2016/02/25/public-support-for-environmental-regulations-varies-by-state/; Gallup, https://news.gallup.com/poll/1615/environment.aspx.

10. https://thehill.com/opinion/energy-environment/370334-epas-budget-has-been-devastated-for-decades-heres-the-math.

of ruining someone's life. This is a large disjuncture between perception and actuality. Laws work better when we don't have that.

After leaving EPA I applied for and received a grant to develop and deliver training on the law. Realtors have to take continuing education, so why not offer a course on lead disclosure that gives them credits toward their requirements? (Thanks to former EPA supervisor Deborah Brown for the idea.) This can be done all across the country, and I created the course in hopes that EPA would pay for that to happen, and they paid me to do it in New England for six years. But that money has gone away and no national effort was launched.

My impression from giving the class to a few thousand attendees is that-many took it for the credits, thinking also that it would help them stay out of trouble, and expected it to be incredibly boring. They had heard about potential large fines for violations. But judging by the response to my training, I am convinced that some also want to protect children from unnecessary harm, and that many more acquire that goal once they understand the problem a bit better. They seemed moved by learning how bad lead is. Showing them pictures of how brains have been damaged by lead poisoning is one way of getting the point across. The overwhelming response to the course has been positive. Helping the regulated population understand *why* it has requirements helps boost compliance.

For my grantors I conducted before and after surveys so I could report on results. Attendees unfailingly indicated a strong shift toward increased understanding of the issue, its importance, and commitment to change. Education works.

But what I learned is that the law is not working, very much, as intended. It has some effect, but it could have much more if people "got it." Few do.

First of all, the rule is very short, just a few pages, but almost no one had ever read it. The Lead Disclosure Rule[11] says that sellers or lessors of residential properties built before 1978 must give out a pamphlet and disclosure form that warns that lead might be present on the property. Maybe a quarter of the audience would express irritation at the bother of this law. I would ask, "Why is it so hard just to give over information?" They would say: *because drawing attention to the issue of lead would lead to all kinds of trouble. They would be made to spend money to remove the paint. They would be sued.* But this was not the truth of the matter. I would draw their attention to this sentence in the rules:

> Nothing in this section implies a positive obligation on the seller or lessor to conduct any evaluation or reduction activities.

11. 40 CFR 745, Subpart F, "Disclosure of Known Lead-Based Paint and/or Lead-Based Paint Hazards Upon Sale or Lease of Residential Property."

The rule is just about disclosure. It says there it *doesn't make you test for or remove the paint*. And if people are aware of the lead in the house, they will be more likely to keep it clean or encapsulate it, in order to live safely, and not have to sue or report anyone. Raising consciousness about lead will increase the chances people live with it safely, and that lessens the chance they have any reason to want to sue you.

This was virtually always a new thought. Complying with the intent of the law—its spirit as well its letter—that tenants and purchasers be thinking about the lead that might be there, is actually good for the landlord and the seller. It seems counterintuitive to some. But people sue because they've been harmed, so reducing the chance that they will be harmed reduces the chance of suit.

I would also note, for the Realtors, that the law requires disclosing records or reports "pertaining to" lead, not just proof that lead is there. They all seem to think that if they don't test for lead, that they then have nothing to disclose. So they don't test. EPA advises real estate professionals to test, but *the regulation as the regulated population understands it creates a disincentive to test.*

EPA has clarified that it cares about the responsibility to transmit information you have about lead, even if it is uncertain that it even means lead is present. It published *Interpretive Guidance*, which none of my course attendees had ever heard of, much less looked at. Outreach is needed to educate them about how EPA interprets the law, because until a court says otherwise, that's how the law is enforced. But just publishing *Interpretive Guidance* on the web is not enough to get the regulated community to know what they need to know. An agency without resources, however, can only do so much. Question 40 (published in 1996) asks this:

> EPA's pamphlet "Protect Your Family from Lead in Your Home" states that recent studies suggest that home test kits for lead are not always accurate, and that consumers should not rely on home test kits to assure safety. Therefore, does the use of home test kits for lead constitute knowledge of lead-based paint for disclosure purposes?

It answers:

> Yes. If an owner has information obtained from the use of a home test kit for lead, that information must be disclosed; however, the owner should also disclose information about the reliability of the test kit results.[12]

---

12. *Interpretive Guidance for the Real Estate Community on the Requirements for Disclosure of Information Concerning Lead-Based Paint in Housing, Part II*, December 5, 1996, Office of Pollution Prevention and Toxics, U.S. Environmental Protection Agency, and Office of Lead Hazard Control U.S. Department of Housing and Urban Development, pages 6-7, https://www.epa.gov/sites/default/files/documents/igd.pdf

This means you don't only disclose *proof.* You disclose information pertaining to lead.

Question 41 is also interesting. It asks, "Must records/reports involving lead-based paint or lead-based paint hazards that *no longer exist* be disclosed?" (emphasis added). The answer, which is yes, includes this:

> Lessors and sellers are obligated to disclose all known information, including information which shows that the lead-based paint or lead-based paint hazards have been corrected.

There it is again. All known information. Everyone has heard the phrase "ignorance of the law is no excuse." Knowledge does not only mean what you happen to know, but *includes what you should have known.* So refusing to accept information offered to you (such as a lead test by someone declining to buy), not looking through your files for information, closing your ears when Uncle Joe says, "I painted that place with lead paint because it's the most durable" may very well not work.

When every Realtor fills out the disclosure form, there is a line called the "agent assurance." Realtors rarely know what they are signing.

The answer is in Section 745.115 of the rules, which I always hand out so they can see for themselves. It says:

> a. Each agent shall ensure compliance with all requirements of this subpart. To ensure compliance, the agent shall:
>
> 1. Inform the seller or lessor of his/her obligations under §§ 745.107, 745.110, and 745.113.
> 2. Ensure that the seller or lessor has performed all activities required under §§ 745.107, 745.110, and 745.113, or personally ensure compliance with the requirements of §§ 745.107, 745.110, and 745.113.
>
> b. If the agent has complied with paragraph (a)(1) of this section, the agent shall not be liable for the failure to disclose to a purchaser or lessee the presence of lead-based paint and/or lead-based paint hazards known by a seller or lessor but not disclosed to the agent.

All those sections referred to are the sections that describe the disclosure requirements. Note how firmly it states that each agent shall ensure compliance, and that the agent must "personally" ensure compliance. To understand what this means, I do a thought experiment with them.

Imagine you are sitting in the office of your client, and they have a report on lead in the residence on the corner of the desk, and you can plainly see that's what it is, but they say to you that they have no knowledge of lead in the residence. Does this mean you have to tell the prospective purchaser yourself? I don't know the answer but I presume it does. I leave this to them to mull over. Many volunteer that they would. Most say that they would make sure the client knows that it has to be disclosed. I tell them, "That's what you are promising when you sign the agent assurance." Most of them go, "Oh!" It's suddenly clear now. They'd been signing it for years without thinking about it. It says, inform the seller of the obligations under all those sections—they'd never looked that up—they didn't know what it meant. They just signed it.

I end the class by going to the enforcement section and going over how huge the fines can be, and how they can be avoided so simply and reduced through sincere good faith efforts. I make sure they get that knowledge includes "should have known," and that it says there that there will be "joint and several liability" for violations of the law that cause harm.

Joint and several liability. What's that, I ask them. It means you and your client who didn't want to reveal information are both liable, and what's more, a plaintiff can go after just you for compensation. Then I return to section 115 and look at (b). Let's go over it again. If the agent has informed the seller or lessor of their obligations, then they *will not be liable*. I stop there for a moment. Wait a minute, you mean if they had not informed the client about how they really have to take the disclosure rule seriously, about how they need to look in their records, consider what they know, that it's not proof but any information, they c*ould* be liable? I let that sink in and then we talk about the liability to which this refers:

> the failure to disclose to a purchaser or lessee the presence of lead-based paint and/or lead-based paint hazards known by a seller or lessor but not disclosed to the agent.

This information settles in with a kind of shock. The agents realize that even if a client—I call him or her the "evil client"—has information about lead that they don't tell the agent about—that the agent can be liable. They look up and say, "But that's not fair! How can I be liable for something I didn't know about?"

I can try to explain why this is fair relative to the family with young children vulnerable to lead paint. I explain that this awful potential liability is intended to make certain that they will inform the client of the need to disclose any information pertaining to lead—(not proof)—that is known to them. This may not be convincing, and it does seem to go against a fundamental principle. But whether or not you agree, the point is this: they can escape that terrifying liability by making sure they inform the client of their responsibilities to disclose.

But the law doesn't work the way it should if people don't understand how it works! The purpose of these provisions was to put the fear of God into

Realtors so they would make sure clients got the point and did disclosure. But agents who took this class did not know they faced this form of liability or had this means of escape from it. They sign the agent assurance line without thinking. And clients, from what I'm told by my class attendees, generally do not look in their files, generally do not test for lead, and generally think they are better off playing this whole thing down. And customers often receive the lead pamphlet without reading it and learning what they need to learn.

Though the numbers (in New Hampshire) are growing, it is still a distinct minority that realizes that raising attention to the possible or actual presence of lead in a residence lowers the risk that occupants will be harmed by it, for they are more likely to take the necessary precautions. This is a story about a law that works through awareness with a level too low to work as intended.

Now, after years of delivering this training in New Hampshire, now when I ask if people there provide the pamphlet, everyone does. But when I ask if they ever open it up and discuss what's in it, only some do. Some open it up to show families with children the page that says wash your children's hands often, wash the windowsills, give them foods rich in calcium to protect them from the uptake of lead, and for do-it-yourselfers, the page that says how to work without generating lead dusts. But when I began these classes, many did not answer that they provided the pamphlet (huge fine if you're caught), and none answered that they ever opened it up or discussed it. The pamphlet with all this valuable information to keep people safe would just be in a pile of papers, unread, unused, not at all effective in reducing the risk of lead poisoning.

The outreach has clearly improved things.[13] But I was just one person, working only when a real estate group was willing to pay me. It was not good enough for Congress to just pass this law, and then fail to give EPA enough funding so that it could hire one hundred people like me to do this kind of training in every state. (When you provide free or discounted training, real estate people come for the continuing education credits.) Without that engagement, the law simply does not work as intended. It is delusional to think that simply creating legislation is enough, simply promulgating regulations are enough. *Substantial engagement with the regulated community is required for our laws to be effective.*

When I worked at EPA, there was an effort to do outreach. It consisted of telling people what the law required. It was a good thing, and it helped. But the kind of outreach I'm talking about includes asking questions about what

---

13. See the final grant report to EPA, in 2010, "Engagement with the Source," at http://www.bu.edu/rccp/files/2014/04/Read-the-reports-of-the-RCCP..pdf. The project reached an estimated three thousand real estate professionals throughout New England, "resulting in a change from less than one third considering it of high importance to an overwhelming majority considering lead as of high importance" (8.)

people think and do. Without doing that, and building trust so that people answered my questions, I would not have learned how poorly this law is working.

I stress the example of lead because we have known about its dangers since antiquity.[14] It's not something we can say, well, we didn't know how bad it is. We know. So making sure everyone understands the law and why it's important should be a priority action. But we ought not to have lead in our environment at all. Lead does not break down and it should never have been put into paint or other items used by people. As with all persistent poisons, we should control uses that result in dispersion or exposure. For example, if you put lead into a computer, and all computers were collected for proper recycling, then what's the problem? You don't suffer exposure when using the computer. But because we don't make sure all computers are collected and properly recycled, we get lead exposures when they are shipped to less developed countries to be recycled in the crudest way, or just dumped. There's no excuse for our failure to act to protect people from lead, or our failure to ensure that the laws we have passed for that purpose work as intended.

This is true of many toxic chemicals. Heavy metals are elements, and do not break down into something harmless. They are like persistent, "forever" chemicals. Many organic chemicals have serious impacts on many organs, as does lead, and safer alternatives exist for many, if not all, of these dangerous substances. The example above of a disclosure requirement for lead should apply to many other chemicals we allow in commerce, and our approach to regulation must extend beyond disclosure, for even if the rule worked as intended, it would not afford sufficient protection. Some things should simply be left in the ground or the laboratory. Understanding of this basic principle is lacking. The quality of our awareness of how to provision our world is so low we don't realize we should take action to prevent the introduction of permanent poisons into our environment, into our bodies. We need to envision a future in which there is greater understanding of such important concepts.

TORT REFORM

As with all problems no analysis is complete without getting to the root of the problem. We tend to choose the least socially disruptive solutions, but this is an inappropriate focus on the immediate—akin to being intellectually

14. "Lead's toxicity was recognized and recorded as early as 2000 BC," Dr. Herbert L. Needleman, "The History of Lead Poisoning in the World.". Needleman's work proving the impact of lead on IQ was key to modern lead law (https://www.lead.org.au/history_of_lead_poisoning_in_the_world.htm#:~:text=Hippocrates%20related%20gout%20to%20the%20food%20and%20wine,,be%20a%20result%20of%20the%20enormous%20lead%20intake.

lazy and morally bankrupt, because it ignores eventual injustice. There is too much avoidance of the possibility of disrupting investments. We see this in the failure to ban the incorporation of such harmful substances in articles placed in commerce, and in the current nature of tort law.

As discussed in the chapter recommending anticipatory governance, American courts did not fully apply strict liability, the doctrine of responsibility for harm when it is under your control, as in the English case *Rylands v. Fletcher*; we only apply it to "abnormally dangerous" activities, such as explosives or radioactives.[15] Though lead is extremely dangerous, it is considered normal, part of ordinary experience, and except where some states have acted concerning paint, we do not impose strict liability for harm from its use. That means that if you are harmed by a product that contains lead, you have to prove the maker or seller was somehow negligent—that they departed from what a reasonable person would do. If the product is considered normal and ordinary, then what's the departure?

Let's say that you buy a highly decorated platter and nothing on it tells you that the glaze is leaded, and you use it to serve sliced fruit to your children, every night of their lives. The fruit is acidic and it causes the lead to leach out of the glaze, and before long, your children have high amounts, which isn't even discovered for a long time, because testing is generally recommended for very young children and isn't necessarily done for older children. It has also caused the adults to have high levels of lead. The lead poisoning of the children affects their performance in school and their ability to control their emotions, and gives them frequent headaches and stomach-aches, and in the adults, it increases their loss of temper, reduces their ability to exercise good judgment, and gives them high blood pressure as well as other serious physical and mental health problems. The family is seriously impacted. Let's say that they are lucky enough to discover the cause, after testing the paint on their walls, their water pipes, and the soil outside, and they put those pretty platters on the wall as decorations, and no longer eat off them. But what about their medical bills, their pain and suffering, the expense of the tutoring and therapy that the children need to stay on track, the reduction in earning power of the children because, when they grow up, they will not have the capacity to succeed that they might have had?

In order to get compensation, they will have to find the maker of those platters and prove that they did something wrong—for example, failed to warn of the lead. They will have to prove the maker—maybe the seller too—knew or should have known that the lead was in the glaze, and that they made and/or sold the product knowing that and failed to tell anyone. Then they will have to

15. There are other forms of strict liability, meaning fault does not have to be proven, applied in product liability cases, to the cleanup of contaminated sites, and to permit violations.

prove that the lead—that particular exposure and not another leaded product or another thing that causes those problems—was the reason for their suffering. This is a lot to prove. Let's say they don't have money to pay a lawyer by the hour but need to find a lawyer who will work on contingency—who will only get paid if he or she wins. That's a big risk to take, and the lawyer will need to fund the high fees of expert witnesses to testify to all those facts.

Here we have an example of the law being blind—not because of impartiality—but because it is not looking where it needs to look so that we have a just society. We can change the law through statute. Starting from the viewpoint of what's fair, we realize that people should not be placing such constituents as poisonous, persistent lead into products that people use. We can easily see that product makers should know that this will lead to harm. The harm is *foreseeable*. Thus, responsibility of some kind should attach.

If we are not ready to take the step of banning lead from uses that cause exposures, limiting lead to utilization where it always remains enclosed, then we should at least take the step of imposing liability on those who *recklessly place it in commerce*. In 2019, I was introduced to Massachusetts State Senator Pat Jehlen, who agreed to propose a bill that I wrote to do this.[16] The bill did not gain enough support to become law but it has just been refiled. Bill SD. 1492, among other things, states that:

> the inability to identify the specific party that placed lead or lead-containing products into commerce shall not prevent those harmed by such placement from recovering for damages caused by exposure to the lead or the lead-containing product. When a plaintiff is shown to have lead in blood, bones, teeth or other body tissues at levels identified in the scientific literature causative of injury such injury shall be presumed to have been caused by lead exposure. Plaintiffs suffering injury consistent with exposure to lead caused by the placement of such products in commerce may recover damages for their injury by establishing by a preponderance of evidence the following:
>
> (1) that the defendant placed lead or lead-containing products into commerce which caused exposures sufficient to cause injury, and (2) that the party placing lead in commerce knew or should have known that such harm was a likely consequence of their actions, and (3) the party placing lead in commerce took no or insufficient action to prevent such harm.[17]

---

16. Thanks to Rick Rabin for the introduction and help, as well as Eugene Benson, Laura Maslow-Armand, and Mark Martinez.

17. https://malegislature.gov/Bills/192/SD1492. See a short video made by Kai Medina when he was a student of mine, describing the bill: https://www.youtube.com/watch?v=AxronhtudPU. I am grateful to Erik Olson of the Natural Resources Defense Council for the suggestion of including a presumption of causation when injuries are consistent with our scientific understanding of what lead poisoning does. To see a recent conference on lead with presentations from experts showing the range of sources we need to address, and steps we can take to eliminate lead poisoning, including a presentation by Senator Jehlen on the bill, click on "Events" at www.bu.edu/rccp.

The bill says if you can show you took some action, or there was no reason to think the product would be used to cause exposures, then you would not be liable. It focuses on manufacturers and doesn't provide for suit against landlords, guardians, schools, retailers. They may of course be negligent under other theories, but thus far they have been the main targets of plaintiffs because of the difficulty of reaching the primary cause—those who should not have put the lead in products in the first place.

The establishment of a presumption that lead caused harm that is the kind of thing we know lead causes shifts the burden of proof, from the plaintiff to show that lead was the cause, to the defendant to show that lead was not the cause. Again, start from what's fair. Is it not unfair that innocent victims should have that burden? Or more fair that it be placed on those who put lead in products? In addition, which policy better serves the general public interest in reducing the amount of toxics we add to our environment?

The bill also says that the judge can use any reasonable means of allocating liability amongst the defendants. Let's say you have thirty-five makers of the poisonous platters. You found one and sued them. Under our current system of law, they can say, "Prove it was me, and not one of those other platter-makers. Show us your receipts!" But you bought those platters many years ago, and you don't have receipts. Should you be without a remedy? We don't see many courts finding ways to dispense the justice needed. It is clear that the maker was part of a larger group that caused the harm. The bill does not specify how liability should be allocated, but by legislating that courts can do this, it seeks to prompt the development of reasonable methods.

The conservative movement has its ALEC, the American Legislative Exchange Council, which promotes model legislation, such as the Regulatory Review and Rescission Act,[18] that would give governors the power to rescind regulations that don't pass a cost-benefit test. This violates the separation of powers principle, where it is the role of the legislature to decide how government should invest its efforts, not the governor. The governor's role is to carry out—to execute—the will of the people, as reflected in the legislature, not negate it. Cost-benefit analysis is heavily weighted toward the interests of the regulated community, as their costs in compliance are easy to measure and immediate. These costs also often diminish considerably after regulations go into effect and people adjust to them. But cost-benefit analysis does a very poor job of evaluating impacts. In the example above, the company putting lead into its platters will have an interest in claiming that it's very burdensome to them to find a substitute for lead that works as well, and lobbyists can quickly come up with claims of high costs to the industry. But the families who will be harmed—they are in the future, they don't know they will be

18. https://www.alec.org/model-policy/regulatory-review-and-rescission-act/.

harmed, and our science of predicting is not very good. What's more, imagine that harm as something that is moderate and spread among many. Imagine that millions have an increase in stomachaches and a decrease in IQ and emotional control, but none of it rising to the level of being something we notice enough to track to its cause. Our estimate of the benefits of not having lead may not be calculable, and they may seem small, because they are so widely dispersed. They may seem ordinary. But if the legislature decides that lead should be eliminated, on principle, this law would not let that stand unless you could prove the benefits were greater than the costs. That is the work that ALEC is doing across this country, benefitting chemical and fossil fuel and other corporate interests who have turned cost benefit analysis into a tool for slowing environmental policy progress. But just as ALEC can promote such a law to state legislatures, advocates for a less toxic world can promote laws to shift the burden to those who create harms, who should know that they ought not to be placing persistent poisons in the products we use, no matter what the tally of costs or benefits. The state legislatures belong to the people if the people use them. They can use them to stop the continuous, slow drip of poison into our lives.

EXERCISE

Think about what's fair and right, and find a legislator who will listen to you. Sit down with a couple of people who know what they're doing, and write up a bill. Maybe this will become law someday.

More people can become involved in making new law happen. We don't have to stick with the old, outmoded, ancient system of tort law that we have inherited that is manifestly unfair. We can envision what's fair and design something new. If makers of things know that they might have to compensate those injured by their truly negligent actions, they will be less likely to perform them. ALEC has no compunction about proposing changes to law that benefit its well-funded sponsors.[19] We should counter its efforts with proposals that will benefit ordinary people.

Now the knee-jerk reaction to this is that it will unleash a horrible storm of litigation and no one will want to make anything anymore. But hold on a second. If this law passes and people think twice about using lead, will that be what happens? Not likely. Most likely is that they will be discouraged from using lead, or will use it only in ways not likely to harm people, as the bill allows. Manufacturing of safer products will be the result. It is true that a poorly written law could have such terrible effects as discouraging beneficial

19. https://www.commoncause.org/democracy-wire/who-still-funds-alec/.

manufacturing. But good-faith open discussion of policies that will bring about fairer law ought to be the remedy to this potential problem, not shying away from trying to reform an unfair system that allows such reckless imposition of risks. We could implement this law unfairly by having a bad application of the concept of recklessness, for example, but that need not happen if we openly discuss this possibility and work to avoid it.

Another approach to the issue would recognize that government has a duty to act to protect its citizens if it can, and here it can provide analytical services. Ordinary people are not going to go out and get a device to test their platters or pipes or imported toys or spices to see if they have lead, but government can provide such services, at a location where people can bring the items to be scanned, or as part of inspectional programs, and by scaling up product testing (particularly of imports).

The quality of our awareness should include not just understanding the problem but all of our options for addressing it. We can profit by spending time envisioning a future tort system that is fairer to victims and more effective at deterring things that should be deterred. We should also consider better ways to help victims than having them have to sue, and that can be achieved through legislation to establish compensation programs, paid for by a fee levied on those who profit from creating the risk.

EXPANDING THE FOCUS

If you pay attention to the issue of lead poisoning, you cannot escape recognition that its impact is greatest on those who are least powerful—the poor, people of color—(who in Flint, Michigan, were given poisoned waters after their democratically elected government was deprived of authority by the appointment of a city manager), and who have for more than a century lived in homes with peeling lead-based paint, and had to bear the idea that lead poisoning was simply what happens to the poor.[20] We still see, if we choose to look, the reluctance to see the harm still being inflicted, in poor water infrastructure in thousands of cities, as, for example, Jackson, Mississippi, whose residents must boil their tap water, if they received it, after the deep freeze burst water pipes throughout areas of the south. Jackson, like Flint and many other cities, has experienced an eroding tax base as white residents and

---

20. At the 1957 annual meeting of the Lead Industries Association Manfred Bowditch argued that "'the major source of trouble is the flaking of lead paint in the ancient slum dwellings of our older cities'—though in saying this he obscured the fact that lead had been a major component of the paint applied to buildings as recently as a decade before. 'The problem of lead poisoning in children will be with us for as long as there are slums,' he argued," Markowitz and Rosner, *Deceit and Denial: The Deadly Politics of Industrial Pollution* (2002) 103.

businesses left the area, and neither state nor federal help arrived to assure access to basic needs, such as safe drinking water. According to the *New York Times*, Mississippi Governor Tate Reeves has "hinted that the state might take over Jackson's water system," while the city's Mayor Chokwe Lumumba seeks state funding.[21] As Northeastern Law Professor Martha Davis writes in "Let Justice Roll Down: A Case Study of the Legal Infrastructure for Water Equality and Affordability,"[22] research is establishing the fact that "US water and sanitation inequalities occur along lines of race, ethnicity, or gender, or disparately affect people with disabilities." First, we need to see this problem, and second, we need to see that the rising cost of water is worsening the situation, and third, we need to see that having safe drinking water is a basic human right. Then perhaps we can see the need for *a systemic change*, from one that expects local property taxes to provide such basic necessities, to one that addresses the needs of all citizens, without regard to whether their municipalities are solvent.

CONTROLLING THE SOURCE

After lead was taken out of gasoline, residential paints, tin cans, and water pipes, we saw a tremendous reduction in lead in people's bodies, through controlling the source, not trying to mop up the mess after it happens.[23] Forty-five years ago, people had an average level of nearly thirteen micrograms per deciliter in their blood, and today it's less than one. But lead is still on the walls, in pipes, in soil, and *still even being placed in products*. It is still dispersed into the atmosphere from small aircraft, and while EPA announced on January 12, it will take action on that, the announcement stated:

> it will evaluate whether emissions from piston-engine aircraft operating on leaded fuel contribute to air pollution that endangers public health and welfare.[24]

There is no question that adding a serious poison that does not break down to our atmosphere is endangering public health and welfare. Scientists agree that no amount of additional lead exposure is good. So the question ought to be

21. Ellen Fentress and Richard Fausset, "In Mississippi, Failing Water System Is Linked to a Cascade of Problems,"*New York Times*, March 13, 2021.

22. *Georgetown Journal on Poverty Law and Policy* 23, no. 3 (Spring 2016).

23. Ninety-three percent from the 1970s. Dignam et al., "Control of Lead Sources in the United States, 1970–2017: Public Health Progress and Current Challenges to Eliminating Lead Exposure," *Journal of Public Health Management and Practice* 25 (January–February 2019) , https://www.ncbi.nlm.nih.gov/pmc/articles/PMC6522252/.

24. https://www.epa.gov/newsreleases/epa-evaluate-whether-lead-emissions-piston-engine-aircraft-endanger-human-health-and.

stated as whether it rises to a "danger" level, and that ought not to be the primary question. It's clear we need to eliminate lead where we can, and that it's only one of many poisons, so we should be constantly looking for the ways to reduce the level of insult to our bodies that we are experiencing. We should also reduce the level of insult to our intelligence by discarding the approach of deciding first whether we have absolute certainty that we are in danger (as if that is the only reason to take action). Endangerment findings should trigger a kind of action, a more aggressive action, but when government also acts to capture and develop opportunities to prevent problems, we have a far more efficient societal response, and a chance to solve them once and for all.

ADDRESSING PROBLEMS CAUSED BY ADDRESSING PROBLEMS

A deeper look would also make clear that our solutions sometimes create ancillary problems. In the case of lead, the Massachusetts requirement that rentals to families with children be made safe has created an incentive for discriminating against families with children. The problem has gotten so bad that advocates for family housing have sued the state to eliminate the provision of the lead law that pertains to children.[25]

Though current anti-discrimination law does not work as intended, and removing the specific reference to children would make the law pertain to everyone, theoretically bringing about universal lead safety, this could be a jarring and too-rapid adjustment for the real estate market to make. It would be better to plan the new requirement of lead safety for all so the transition can be manageable. Philadelphia has moved to do this.[26] Universal lead safety is feasible, and clearly necessary, as it is not just children who are vulnerable, but teenagers and adults as well.

We have focused on children because the damage to them is most severe, but that's not a reason to neglect the need to protect everyone. It is not crushingly expensive to cover up the lead with encapsulant paints, that seal it in, and to remove the windows, which are the main source of lead dusts. Replacing old windows with new ones enables us to increase energy efficiency, which will produce savings over time. The Centers for Disease Control estimates that it would cost $80 billion to prevent all children in the United States from having any detectable lead in their blood, and that *taking this*

25. https://www.massfairhousing.org/uploads/newsletter/Press-Release-11.22.19.pdf.
26. Colleen Macauley of Children First PA explains the law in "Eliminating Lead Poisoning in 2022," at "Events" at www.bu.edu/rccp.

action would bring savings in excess of the cost.[27] Others have come up with far higher estimates of the value of preventing lead.

It is impossible to put a figure on the value of the meaning of such an investment. We don't need to crunch numbers. What we need is to emphasize that all children—of all ages—matter. The quality of our awareness should be measured by our ability to see the value in systems that respect citizens, no matter where they are, and no matter who they are. The quality of our awareness should be measured by our ability to see the big picture and to value who we might become, and whether we act as we could to prompt the evolution toward freedom from pollution and poisoning that we need.

THE GLASS-TO-GLASS RULE

I was a participant in the four-year Common Sense Initiative of Carol Browner, EPA administrator during the 1990s, who implemented Al Gore's vision of reinventing regulation. I found it to be a demonstration of the potential for progress when constructive ground rules are set for seeking consensus among the many relevant perspectives needed for healthy democratic deliberation. But this sort of project cannot be undertaken without staff, and the effort was very much slowed by a lack of adequate staff engagement.

One project in which I was involved had to do with examining how we could craft a new regulatory pathway for recycling glass from old computers and televisions. Old TV and computer monitors could contain as much as eight pounds of lead. In explaining this, I would often recall that when I was a child we were constantly told not to sit so close to the television. We didn't understand why, but the fact is that the images on the TV screen were created by a ray gun firing electrons at the back of the screen, the energy of which would travel through the screen and irradiate our bodies. Lead incorporated into the glass provided protection and made the glass easier to work with.

Today we have a different technology and there is no need for the cathode-ray tubes (CRTs) that once enlivened our days by firing directly at us. At the time of the Common Sense Initiative, however, this liquid crystal display (LCD) was not in wide use yet, and companies making CRT monitors for computers and televisions wanted to know why they needed to overcome very difficult regulatory hurdles in order to use old glass to make new screens. The problem was that lead in the glass made it a hazardous waste. Technically. For all obvious purposes, it was still just a big hunk of glass with the lead safely in it. But once you start breaking up an old TV and treating it as a waste, the lead begins to be released.

27. https://www.cdc.gov/nceh/lead/prevention/default.htm.

The environmental and business viewpoints coincided, in committee, in a consensus that we should create an easy regulatory pathway for leaded glass to be re-melted into new TV or computer glass. We called it the "glass-to-glass" rule. It seemed a no-brainer. Unlike hazardous waste liquids in a drum, there was not much risk of leaking or spilling. You could stand next to or touch the glass and not be harmed. Until the glass was broken up, it was safe, and if you could remake the old leaded glass into new monitor glass, you would reduce the need to mine and transport lead in its pure form. By reducing manufacturing costs, you would help save US companies and jobs. It was a win-win-win. But it took years for the glass-to-glass rule to be promulgated. We did our work in 1998 and 1999, and it wasn't until 2006 that EPA changed its regulations so that CRT glass properly recycled would not be considered a solid waste.[28] If it's not a waste, it's not covered by strict hazardous waste rules.

You can say well, that's only eight years after we first identified the problem and proposed a "conditional exemption" from the onerous rules that were inhibiting this material recovery. In the scheme of things, once you have become accustomed to very long delays or never being able to fix regulatory problems, that's a big success. At least it happened! However, the advent of the new screen technology made this regulatory fix very much less relevant than before. By 2016, CRT glass had essentially lost its market value.[29] The rule came along in time to help to some extent, but it would have helped vastly more had the rule been swiftly promulgated. Why did it take so long?

I can only report what I saw. All during the Common Sense Initiative, we had a constantly shifting group of EPA staff working with us. The main facilitators were always there, but for an initiative like crafting a new rule, we needed EPA attorneys and hazardous waste experts. We rarely had the same staff at two meetings in a row. They were constantly being pulled to deal with suits against the EPA, to help administrators prepare for hostile questioning by Republicans in Congress, to put out fires, and comply with court orders. EPA was simply far too short on staff to provide consistent participation by the needed experts. As a result, we found ourselves briefing new people at the beginning of every meeting. They had to overcome the suspicion that all of this was a trick, or that we were missing something. They were doing the right thing, checking things out very thoroughly before getting on board, and they always ended up endorsing and assisting the effort. But this took up time, and we were always starting over from the beginning.

---

28. *Federal Register* 71 (July 28, 2006): 42928.
29. https://earth911.com/business-policy/whats-blocking-crt-recycling/.

Then, when the rule was pretty much written, George W. Bush was elected, and his administration seemed to care nothing about what the Common Sense Initiative had produced.[30] I do not have inside knowledge of what happened next with the glass-to-glass rule, but I do know that it was not long before Al Gore's list of identified opportunities for improving regulation had been taken down from the web, and it was not long after that that I contacted EPA to obtain documents from the Common Sense Initiative for an article I was writing, and was told that the CSI boxes I wanted had been thrown away.[31] I am grateful that people at EPA continued to work on the rule and that it did finally come to fruition, and expect that this demonstrates the value of mid-level staff and how they provide continuity, stability, and deserve far more power and respect within the system, to balance the distortions and instability caused by some political appointees.

The glass-to-glass rule is a story with two sides: one is that it could have made a real difference had it emerged much earlier, and it shows the need for more staff and such engagement by the government so that it can fix its own programs; but on the other hand, it is a demonstration that our hazardous waste law can be improved—it can be reshaped to encourage more recovery of materials, and not just route everything into a black hole of problematic material.

CONFUSING PRESENTATION OF LEGAL INFORMATION

As an example of how communication is hindered by a shortage of staff, I point to the Administrative Procedure Act of 1946, which requires that agencies provide notice of new, proposed regulations, and an opportunity to comment on them, before issuing them as final. For years, I have provided training in environmental law for many kinds of businesses. When people complain that rules don't make sense, I tell them that this law gives anyone the chance to tell EPA about what does or doesn't make sense in a proposed rule. Most people are pleased to learn that we have such a government, and some are surprised that it is designed to hear from them. It helps to jar some people out of a sense that they are regarded as unimportant, that they are not connected to a system of government, that it is distant and could care less about their opinion.

But then once you have introduced people to the Federal Register (FR), where proposed regulations are posted, you have to tell them a few crucial

30. "Reinventing Government—Two Decades Later," *Government Executive*, April 26, 2013, https://www.govexec.com/management/2013/04/what-reinvention-wrought/62836/.

31. I was told this by one of the DFOs (the designated federal officers for the Common Sense Initiative).

things, *or your work to convince them that we live in a real democracy will be for naught.*

You have to tell them that if you want to know what the proposed rule looks like, go to the end of the Federal Register Notice. For first will come a statement of authority, and all sorts of front matter, explaining why the agency is issuing a new rule. This can go on for many pages. When I tell this story, I get some nods, because at this point some people remember their experiences trying to understand the law. On the EPA website, it says if you want to read the law, "go here." This happens in several places. Sometimes it takes you to the Code for Federal Regulations (CFR). If that's the link you clicked on, you are in luck. That is where the regulation resides after the end of the year in which it was promulgated. You will just see the rules themselves. But if the link takes you to the FR, you will see all that front matter, including a summary of what happened after notice and comment, and that will be even longer than the proposed rule, for it will contain a discussion of what people said and why the agency went along with their suggestions or did not. When I mention this, I always get people who remember their frustration at wanting to understand what they are required to do and their eyes glazing over before they can get to the information they want—the rule itself. They just want to know what it is they have to do, and they never get there. It would be easy for the website to explain what people need to know, by thinking about what they want to know. It could tell people a little bit about how the Federal Register notice is set up, making clear that *the rule itself is at the end*, that it comes after explanation of why it is the way it is. Just communicating this would avoid a vast amount of frustration.

With adequate staff, there could be good short summaries of the rule, which are sometimes but not always produced. With adequate staff, there could be people to go to meetings of regulated entities to explain the rules, and with adequate staff, there could be far more effort to consult with the affected populations to encourage comment.

When I talk about the opportunity to comment, I know that people like the idea, but they are far too busy to find out how to learn about it, much less take the time to read the proposed rule and think about it. With adequate staff, agencies could do more publicity, make it easier to contact someone to find out more, and host meetings to explain and discuss.

This problem, of course, is not limited to the EPA. It is generic to all agencies. We live in a democracy that has the foundation for citizen engagement. But without the staff to make it happen, it is not real for the average citizen. It is quite real for high-powered lawyers and well-capitalized corporations and trade associations. It is also important to informed citizens and activists. But the great mass of people who need to be reached are not being reached.

This leaves the great bulk of those who must comply with rules or who need them in place feeling that they are not respected. Although this is a democracy designed for the people, it is generally utilized by those who have the capacity to learn and use it. Far more assistance could be provided, with staff, with initiatives, with funding. For example, imagine if after graduating from college, wondering what job you can find, you could easily get a job helping your community navigate the corridors of government, as a member of a national civic engagement service corps.

While the addition of staff to do outreach, engage the regulated population, and explain the rules is needed, the effort to connect needs to go further. The communication we need must go in both directions. The agencies must decide they need to learn as well, and must treat the regulated population as a source of information for improving the rules. Here those national service jobs for the young community organizers, representatives, advocates, facilitators will be especially valuable as a connection to the actual people of the neighborhoods and populations the government needs to understand. Hiring the effective people into the agencies will help cement those connections.

REMEMBERING THE FORGOTTEN: THE EXAMPLE OF NOISE

Noise is another issue that reveals the low quality of our awareness. We don't want to deal with it. Many do not realize that *we once did.*[32] EPA at its start had a noise program, because Congress included noise in the Clean Air Act. It was correctly seen as a pollutant, contaminating our lives and communities. During the Reagan years, the program was defunded, though it remains on the books. Little effective law now exists to deal with noise, except for local ordinances. These commonly refer to noise as a nuisance problem. While it is true that "nuisance" is a legal term that can encompass serious harms, it is also a word that connotes something that is merely an annoyance. Leaving noise as an issue that each municipality deals with on its own, often not at all, and generally as a matter that seems trivial, means that we are really not dealing with it. EPA, in its first decade, amply showed we can act to reduce noise, but we decided not to. One may wonder if people even know this happened and think of it as an example of low quality of awareness of governance.

New information shows how serious those impacts can be. In addition to hearing loss, tinnitus, and hyperacusis caused by loud noise, chronic low-level noise contributes to cardiovascular disease, cerebrovascular disease, mental disorders, and even early death. The American Public Health Association is

32. See, for example, the 1980 report on what the program achieved during the 1970s: https://quietcommunities.org/wp-content/uploads/2022/01/1980_Noise-Control-Program-Progress-to-Date.pdf.

currently updating its statement on noise to reflect newer information, but its 2013 statement asserted that "reducing environmental noise pollution has the potential to save lives," because of new understanding of systemic effects on organs such as the heart. A 2011 World Health Organization report found more than a million "healthy life years" are being lost in Europe due to traffic noise.[33] Like lead and other neurotoxins, noise does not just cause harm with physical ramifications but also reduces the capacity of children to learn[34] and the ability of all humans to function intellectually: it thus affects the productivity of society and harms us all. Are we capable of seeing this, or will we continue to think of noise as simply an irritation? Can we understand this objectively, as a significant impact, or will we continue to think of it as simply a subjective experience by individuals?

On the side of objectivity—and a different subjectivity—we must now recognize the impact noise has on animals. Our joint subjective experience of having the opportunity to privilege our own experience does nothing to mitigate how other life-forms are experiencing our loudness. Did you ever listen to the *Songs of the Humpback Whale*?[35] When it came out, and Judy Collins then sang with the whales, it was as if a new group had enlarged our cultural celebration. There are thinking creatures in the ocean. Whales are no ignoble beasts, but complex, communicative, beautiful in speech, musical. Perhaps, who knows, tender? Philosophical? It became impossible to just continue killing whales after hearing them.

Can we recognize that noise from sonar and ocean oil exploration and ordinary shipping traffic interferes with their communication and is likely extremely painful? Bats use echolocation to eat mosquitos—we don't want to interfere with that. We may assume we don't want to make it harder for birds to talk to each other.

Biologists have stated that anthropogenic noise is now omnipresent in terrestrial and aquatic environments and should be considered a "major global pollutant."[36] They have documented profound effects on amphibians, arthropods, birds, fish, mammals, molluscs and reptilians. Noise is not just an annoyance. It is actually threatening the ability of many species to persist. It is another factor contributing to the denuding of our landscape, the reduction of life in the ecosystems on which we depend.

---

33. World Health Organization, 2011, Burden of disease from environmental noise: quantification of healthy life years lost in Europe. See http://www.euro.who.int/__data/assets/pdf_file/0008/136466/e94888.pdf.

34. See the work of Arline Bronzaft: https://www.k12dive.com/news/a-less-noisy-classroom-equals-a-better-learning-environment/552358/

35. A 1970 album produced by Roger Payne.

36. "The Effects of Anthropogenic Noise on Animals: A Meta-Analysis," *Biology Letters* 15, no. 11 (2019).

We have two kinds of nuisance in American law: private nuisance, an unreasonable or unlawful use of your property in a way that substantially interferes with somebody else's right to use or enjoy their property, and public nuisance, an interference with the rights of the community. What's unreasonable? What do we say is unlawful? Even if you think your neighbor's loud noise is unreasonable interference, your ability to get relief depends on the subjective take of the community, unless we have set standards that can be measured, and the method of measurement is available. It is useful to know that cell phones now have apps that can measure noise.[37] The quality of our awareness includes our ability to use this more objective information and imagine a system that sets a standard and has enforcement that works. For example, instead of having to have the police show up to address a noise condition, you can use a registered app and send your data in to a system that verifies the source, and a warning goes to the neighbor, not just the company they contracted with. In this era in which gas-powered leaf blowers have become widely used, it is important to innovate until we figure out the best approaches. Even without police visits, many people of goodwill would likely stop or reduce their nuisance-producing activity upon receiving the "warning." By narrowing the target population for real enforcement down to the recalcitrant we make that program more efficient and less antagonizing. Using the police for noise makes us want to avoid enforcing noise rules.

A universal vision would not depend on the capacity of the municipality in which we live. Reviving the national program that was established under the Clean Air Act would affirm the right to be free of this form of harm. More than fifty years ago, society correctly recognized noise as a serious pollutant needing to be reduced, just like sulfur dioxides or lead or other air toxics. We should listen to the biologists and see this issue as a major global pollutant. The world is dangerously contaminated with noise. Can we develop the quality of awareness that enables us to see, hear, and act accordingly?[38]

---

37. Recent understanding has led to an increasing effort to define sound as not just subjective but also to use objective measures based on evidence of harm. Noise is unwanted and/or harmful sound. "A new definition of noise: noise is unwanted and/or harmful sound. Noise is the new 'secondhand smoke.'" Daniel Fink, Proc. Mtgs. Acoust. 39, 050002 (2019); https://asa.scitation.org/doi/10.1121/2.0001186.

38. For more on noise, see the website of Quiet Communities, the conference on Quiet, Clean Land Care at www.bu.edu/rccp under "Events," https://quietcommunities.org/portfolio-items/quiet-conversation/, an October 8, 2021 piece asking how the small engine industry got Congress to extend the preemption of state legislation regulating car manufacturing, to gas-powered leaf blowers, which are harmful to workers and highly polluting.

A too-gentle transition also risks compromising with evil and going at too slow a pace. But it may be that it is a lack of firmness, not the presence of gentleness that causes a slowing down of progress. Firmness comes from the pull of the vision of better relationship and the knowledge that this is right and just and deserved. This principle if universally applied is uniting. If rights are our touchstone, we have the justification to be firm, and if they are universally applied, the chance for consensus. Placing things in context can make the facts speak for themselves.

Many may feel it's better to stick with technological advances, or with greening finance, or advancing green purchasing, or a few changes at law. However, clean technologies and practices may be unused, green financing and purchasing may be grossly inadequate, and the right laws may not even be passed, or obeyed if they do gain passage. We need to strongly assert what's right—as, for example, the young people suing the government for not acting to stop climate change working with attorney Julia Olson and others, who obtained this statement by federal district court judge, Anne Aiken, in 2016:

> I have no doubt that the right to a climate system capable of sustaining human life is fundamental to a free and ordered society. Just as marriage is the "foundation of the family," a stable climate system is quite literally the foundation "of society, without which there would be neither civilization nor progress."[45]

Gentle is not the same as weak. Being committed to quiet transition does not require accepting harm or disrespect. There is nothing in being respectful that prevents strength and determination. John Lewis, for example, an apostle of nonviolence, is considered a pillar of strength for his firm resistance despite the violence visited upon him.

By refusing to oppose the opposition with the same methods they inappropriately use against others, we can transcend the binary choice dynamic and inhabit the true context of the situation in which all sides are considered and the resolution of the highest optimality is the purpose of discussion. Eschewing political combat, winning by any means, is not giving up anything if we strongly assert our actual context of being in this boat together.

---

45. In *Juliana v. US*, the Ninth Circuit Court of Appeals has dismissed the case on separation of powers issues, not the merits of the claim, and the plaintiff's complaint may be amended. See https://www.ourchildrenstrust.org/ and https://static1.squarespace.com/static/571d109b04426270152febe0/t/5824e85e6a49638292ddd1c9/1478813795912/Order+MTD.Aiken.pdf.

WE FACE A PSYCHOSOCIAL PROBLEM

There is no avoiding what should be regarded as a fact: our failure to manage this planet, and ourselves, is a psychosocial problem. We must work on the level of emotion, perception, attitude, habits of thought. A society in which the majority has a sense of responsibility must be our goal, and if we can achieve that, through education, social expectations, understanding, and incentives, then we have a chance at making our aspirations real. The mechanics of governance are secondary to the motivation to use governance. Without the latter, we can design the perfect system to no avail.

It is helpful that relevant experts in thinking are turning their attention to the problem: psychologists, psychoanalysts, and psychiatrists have, in recent years, increasingly taken up the question of how we think about environmental problems. Harold Searles wrote in 1972 about the ecological crisis:

> My hypothesis is that man is hampered in his meeting of this environmental crisis by a severe and pervasive apathy which is based largely upon feelings and attitudes of which he is unconscious.[46]

Paul Slovic told us in 2007 that one problem is when the numbers of people suffering are too great, we fuzz out.[47] The American Psychological Association's Task Force on the Interface between Psychology and Global Climate Change told us that people tend to discount future gains and losses, and people are confused about the relevant facts. But Brian Baird, a psychologist and legislator from Washington, noted in an article on that Task Force report that "The quickest way to reduce our energy consumption is clearly behavioral. . . . It's going to take us a while to get to cellulosic ethanol or fuel cells or whatever . . . but we can change behavior tomorrow."[48]

Renee Lertzman notes Searles's work and argues that

> Attention to unconscious processes, defensive mechanisms such as denial, projection and splitting, and nuanced understandings of anxiety, loss, mourning and grief—can help us bring back into the frame the potential presence of concern, anxiety, worry, fears, desires, aspirations and hopes in how we conceptualize environmental engagement (or its lack).[49]

---

46. H. F. Searles, "Unconscious Processes in Relation to the Environmental Crisis," *Psychoanalytic Review* 59, no. 3 (1972): 361–74..
47. "If I Look at the Mass I Will Never Act: Psychic Numbing and Genocide," *Judgment and Decision Making* 2, no. 2 (April 2007): 79–95.
48. Michael Price, "Changing Minds to Prevent Climate Change," *Monitor on Psychology* 40, no. 9 (2009), https://www.apa.org/monitor/2009/10/climate-change.
49. "The Myth of Apathy: Psychoanalytic Explorations of Environmental Subjectivity," in *Engaging with Climate Change: Psychoanalytic and Interdisciplinary Perspective* (New York: Routledge, 2013) by Sally Weintrobe.

Robert Stolorow advises, concerning "the pervasive evasion of the very real possibility of the doomsday scenario posed by the specter of climate change," and the fact that while there are reasonable things we can do about it, but don't, that we

> shift from a philosophical and psychoanalytic perspective emphasizing the unbearable emotions that would accompany a facing-up to a doomsday scenario with its collapse of meaning. It is from the horror of the doomsday scenario posed by climate change that the minimizers, scoffers, and ridiculers turn away . . . we must face up to our apocalyptic anxiety before it is too late for the survival of future generations. Such facing-up requires that we open a far-reaching emotional dialogue in which the Angst can be collectively held and borne.[50]

Rosemary Randall holds "Carbon Conversations" to "offer an important way of thinking about the problem."[51] Lertzman advocates for an approach that acknowledges and provides space for "contradiction, ambivalence, loss and mourning," in order to avoid

> the simplifying and patronizing tone of many environmental messages and the manic emphasis on 'solutions,' as if the messiness of our situation can be avoided and glossed over. Acknowledgment of the painful dilemmas we grapple with can have a potentially disarming effect, ideally softening defenses and sparking creativity and concern. If we frame our communications presuming a presence of care and creativity, rather than an absence, as is often the case, we may see some radical reframing of public engagement.[52]

Sally Weintrobe's 2021 book on the *Psychological Roots of the Climate Crisis*[53] is powerfully subtitled *Neoliberal Exceptionalism and the Culture of Uncare*. The culture of uncare is the same culture Martha Albertson Fineman is talking about, organized for autonomy, desired by those who have the exceptional idea that they are entitled to more than a just share of the world in which we all exist. Such a culture cannot become sustainable, because when we are not responsible for each other, then no one is taking care of the biosphere itself.

Having a dream of the paradise that we may seek is one thing. Convincing people to walk there with us is another. Understanding why some remain stuck, or even to lash out at us as we invite them to come along, is essential.

---

50. "Death, Afterlife, and Doomsday Scenario," *Psychology Today*, February 18, 2014.
51. Paramaguru, "The Battle Over Global Warming is All in Your Head," *Time, Climate Science*, August 19, 2013.
52. *Engaging with Climate Change*, 130. Lertzman refers to D. W. Winnicott's 1963 article "The Development of the Capacity for Concern," in *The Maturational Processes and the Facilitating Environment*.
53. https://www.sallyweintrobe.com/about-sally/.

Psychologists and psychiatrists may be good at dealing with symptoms, and psychoanalysts may be good at helping you to understand what's going on inside of you, and why you do or don't do things, but typically they are dealing with patients one at a time. The patient, in this case, is all of us and we are presenting an infinite variety of symptoms and underlying causes. *But there is something we all have in common. The specter of the environmental catastrophe is an endangerment of everyone's hopes and dreams*. Certainly, there is a simple story available in which we learn to be heroic, or at least what we can be proud to call human.

There are so many simple stories that dismiss our dreams. *Toms River*, the book by environmental investigative reporter Dan Fagin, lays out the history of the horrific pollution of a section of New Jersey by chemical companies When public criticism of the fouling of local waters first became an issue in 1963, which the companies could not easily dismiss, the general manager of one of the companies told reporters that the company had a payroll of millions of dollars, and added:

> If the responsible people of this area want a prosperous company with continuing growth, we must realize there will be changes in our natural surroundings. Very few of us are willing to live like the Indians, in spite of our idyllic dreams.[54]

In 1963, it seemed feasible only to dump waste chemicals, and a crazy dream to operate without poisoning the neighborhood (including drinking water). We know now that wastes can be managed, and we have a dream that they can be completely eliminated. It is no crazy dream, as the company manager was asserting, but warranted and necessary.

In a basic way more important than knowing is feeling, for it frames and maybe prevents thinking. The awareness of the threat to our lives, to all life, that our lifestyles present, can be so terrifying that the advice of Stolorow and Lertzman, to admit the sad feelings, may be rejected as simply too hard to do, and the rejection may be unconscious with powerful resistance to recognizing that. As Stephanie Brody writes, it is difficult to feel loss "in a landscape devoid of others," and referring to the work of Holocaust survivor and analyst Anna Ornstein, states that "Inevitably it becomes clear that it is not just mourning that is resisted, but mourning in isolation. . . . Alternatively, when we trust others to console us in our pain, we are free to enter the memorial space." She writes:

54. Dan Fagin, *Toms River: A Story of Science and Salvation* (2013), 58.

> It is in the remembering with others that we ride the waves of grief, and feel the pull of the tide that connects us to the sadness and then subsides. Grief is best borne in the company of others.[55]

If you have read this far, surely you feel to some extent that you are in the company of others—can glimpse them around you. Governance will more likely be rebuilt the way it should be if we come together, in awareness of each other, in joint apprehension of our realities. That is the first order of business, creating this mood, this culture, this happening, this movement. A society with enough people of goodwill committed to joint thinking, joint systems transformation, shared process and equitable access can write the road map.

The stories in this book are examples that solutions exist, but they have to be invented anew for each context. The real set of solutions is as yet unwritten, for it will work when it is the product of enough of us to represent the united will of a mature people thinking sensibly. It seems unavoidable that we must begin by acknowledging our collective grief at what has been lost: to regret the horrendous slaughter of whales, to call it inhuman, to be horrified at dead zones, contaminated land, mountaintop removal, neurotoxins in children, dying bees, bleached coral, and rising seas. It is time to recognize the loss of the ability to be proud of ourselves and face that future generations will not respect us unless we bear the necessary recognitions and include them in our thoughts and actions.

But shame is dangerous medicine that must be used only very carefully. The transition we need will require art informed by an empathic understanding of the perspectives of all stakeholders, including those not present or able to articulate.

Our awareness during the day is informed and shaped by the dreams we have had all the nights before, and by momentary and long-lasting daydreams. The battleground is our minds and desires. Our weapons—or tools, let us say—are stories that connect us to our deepest feelings, that revive our full humanity. We must make that painful start to find that good ending. As the great fiction writer Delmore Schwartz wrote, "In Dreams Begin Responsibilities."[56]

---

55. *Entering Night Country*, 151.
56. A short story originally published in 1937 in the first issue of *Partisan Review*. Schwartz took the phrase from an epigraph in a book of poems by W. B. Yeats, who credited it to an "Old Play."

EXERCISE

Imagine that you are visited by someone from another planet, who asks you to explain why humans have dug and drilled into the Earth and extracted poisons and dispersed them so that they weaken the systems of living things, and why they have created poisons in addition to those they have been drawn up from what in myth has been called Hell? How do you explain our behavior? It is not enough to blame it on those who actually did the drilling and digging. The alien does not buy it, regarding each of us as a member of the superorganism that did these things. After your attempt to explain the behavior, imagine that you feel compelled to tell the alien, speaking on behalf of the human race, why you will do better, because you are afraid that your explanation has not convinced the alien not to wipe us all out for the benefit of the galaxy that we must not infect with ourselves. What do you say about how we will do better, sticking to what you think is actually possible?

Chapter Eight

An Equitable System

In the classes that I teach, I give the students the choice of topics for most of their assignments. This is intended as educational for them, for it forces them to stop waiting to have decisions handed to them, and makes them think about what they really care about, what they want to know. This is the key to work satisfaction, I believe, and they do a better job when they have chosen the topic. But I also learn from this, not least in the way I see trends over time. For example, when I started teaching environmental law at Boston University in 2000, no one ever talked about food. But now, at least one-third of the students choose food-related issues.

ENVIRONMENTAL JUSTICE

Another choice that has rapidly grown is environmental justice (EJ)—what should really be called environmental injustice. I see also now some interest in Indigenous rights when it had not come up before. Both of these are complicated, as they trigger an awareness of the vast tragedy of history, and awareness of how that persists into the present, and that triggers guilt and a sense of responsibility so great that even those who are compassionate and mean well don't know what to do. Do we give land back? Do we make reparations? Do we site polluting plants in wealthy neighborhoods? What can we do? The questions are illuminating, and begin the process of crafting solutions. The answers are complex. We should not demand to know the outcome before agreeing to begin the conversation. Rather, we should have faith that facing such issues will bring us closer to resolution than running away from them has done.

What's important is to sit with these complexities and not to discard them because they cause discomfort. Looking for creative solutions helps. For example, there is comanagement of federal and state lands with tribes,[1] and the Department of Interior can receive land in trust for tribes.[2] These are examples of how there are options that are not all or nothing. Imagine that you own land that you want to convert into conservation land, and you need it to be managed so that it remains preserved, or you believe it would be just to return it to descendants of the original inhabitants. It may be that comanagement or trust designation is something people would want to do. They need to know that these options are available.

A similar reorientation from all-or-nothing is the problem of locating industry in areas already overstressed by pollution. A simplistic approach is to restrict new development. But an idea promoted by a former director of the Office of Technical Assistance, Paul Richard, was promoted at a 2000 conference he organized: bring clean technologies to EJ neighborhoods. Don't just keep dirty industries out—work to bring clean industries in, so we have no net loss of jobs.

OTA, as discussed earlier in this book, had by this time worked with hundreds of companies, and had direct knowledge of the growth of safer, cleaner industries. Paul's conference turned the old idea of EJ on its head. EJ should be about developing the economic opportunities in those areas, by providing jobs that don't hurt workers, in facilities that don't contaminate the environments in which people live.

EJ IMPACT STATEMENTS AND INCREASED MONITORING

In the fall of 2016, after I had created a project class at Boston University entitled "Research for Environmental Agencies and Organizations," I was asked by EPA's Deborah Brown to see if the students could write a draft model municipal Environmental Justice ordinance. President Clinton had issued Executive Order 12,898 in 1994, and by 2014, all states had issued some form of one, but a glance around the country showed that a pressing need for progress remained. Maybe, Brown reasoned, we needed to work on the local level. I gave the students some ideas, such as Paul Richard's, but waited to see what they would come up with. I have learned that if you give too much instruction, the students are less likely to be creative.

---

1. https://www.cfc.umt.edu/bolle/files/BadgerTwoMedicineReport.pdf.
2. https://crsreports.congress.gov/product/pdf/R/R46647 and https://www.bia.gov/sites/bia.gov/files/assets/public/raca/handbook/pdf/Acquisition_of_Title_to_Land_Held_in_Fee_or_Restricted_Fee_Status_50_OIMT.pdf.

They came up with several ideas—increasing engagement so that local citizens have more of a voice in any new proposed developments that might add to their existing burden, declaring that the local government has a duty to protect its residents from harm, and expediting approvals for clean development that would provide jobs. But the focus was an idea about how to address cumulative impact. The students had discovered that an advisory council in New Jersey had proposed in 2009 various ways to evaluate cumulative impact, to identify thresholds for action.[3] The work used a study based in Massachusetts[4] by Northeastern Professors Daniel Faber and Eric Krieg, and the Massachusetts DEP had also been working on the issue of cumulative impact, but neither state had yet implemented anything groundbreaking. My students proposed an "EJ impact assessment."

My students and I came up with the idea of putting the burden on project proponents to show they were not increasing any existing burden where populations were already overburdened. We were invited to present our ideas to EPA staff, who saw difficulties in implementing it as written, but appreciated the attention it focused on the important question of how we make EJ policies actually work.[5]

In this class, I propose various projects and let my students choose what they want to work on. In a number of successive semesters, students chose to continue to work on this issue. They engaged in talks with city and state officials, and EPA's Environmental Justice program, as well as activists across the country. We found that EJ was something many people cared about, but there was a sense that progress was stalled. My students were encouraged by many to continue working and came away with a sense that the EJ Impact Assessment was something that could be extremely valuable, if we could continue to refine the idea. After a while, we realized that imposing a special requirement of assessing burden before building in an EJ area would simply stop all development in those areas. So we decided that the EJ impact assessment would have to be applied to all development. This led to the idea that we needed to have monitoring to assess impact, and applying this to all development would be of benefit to all neighborhoods, not just EJ. It was interesting—and, in retrospect, obvious—that a focus on the most vulnerable led to

3. *Strategies for Addressing Cumulative Impacts in Environmental Justice Communities,* Cumulative Impacts Subcommittee, Environmental Justice Advisory Council to the New Jersey Department of Environmental Protection, https://www.nj.gov/dep/ej/docs/ejac_impacts_report200903.pdf.

4. Daniel R. Faber and Eric J. Krieg, "Unequal Exposure to Ecological Hazards: Environmental Injustices in the Commonwealth of Massachusetts," *Environmental Health Perspectives* 110 (2010): 277–88, https://ehp.niehs.nih.gov/doi/10.1289/ehp.02110s2277.

5. Spring 2017 and Fall 2016 projects at www.bu.edu/rccp. Students: Hind Alboom, Kimberly Barzola, Elizabeth Hameeteman, Casey Kelly, Alison LaFrance, Katie Walker, Armina Petrescu-Tudor, and Paige Dolci.

something of benefit to everyone—call it Justice for All.[6] EJ is not something special, on the sidelines, of concern to only a portion of the populace.

The monitoring requirement is simply expressed: as part of the development approval process, an assessment of potential impacts would be required, as in the standard environmental impact statement, but there could be no finding of no significant impact without an assessment of existing impact. For example, if you were building a distribution warehouse that would increase truck traffic, you could expect increased diesel emissions. You would not get approval until there was an assessment of existing air pollution in the areas where these emissions would be experienced, and that assessment would necessarily include monitoring in EJ areas. *No more assuming that the population is already not burdened enough because the data is not showing that they are. Absence of evidence for harm would not effectively be evidence without actual monitoring having taken place.*

I showed the class maps of air pollution monitors set up under the Clean Air Act. They are few and far between. I showed them how the Act works—there is a limit on how much pollution can be emitted, but it is set separately for each pollutant. It also allows some loss of clean air up to that limit, if the increment of increased pollution is not too great. These methods of air pollution control fail to address "hot spots" where people are getting a lot of pollution, especially if the monitors are not nearby and we don't even see what people in a particular neighborhood are experiencing but only get an average for a comparatively large Air Quality Control Region. In addition, many emissions, such as carcinogenic air toxics emitted by diesel engines, are not carefully monitored.

A requirement that project proponents would have to set up monitoring in order to get approval would build out knowledge we don't currently have, and need. We recommended not having companies do this monitoring themselves, but pay a fee as part of the permit process that would fund independent assessment, and that all the measurements would be kept in a publicly available database that would continue accumulating data until we have a far more precise and useful picture of actual conditions than we now have.

A recent report by the Institute for Policy Integrity notes that Clinton's Executive Order requires

> that agencies use available statistical information "to determine whether their programs, policies, and activities have disproportionately high and adverse

6. Kamala Harris and others introduced an Environmental Justice for All Act in 2020, which would give EJ communities a greater voice and require closer consideration of cumulative impacts, including those from climate change, on already burdened communities, https://ehsdailyadvisor.blr.com/2020/08/what-is-the-environmental-justice-for-all-act/.

human health or environmental effects on minority populations and low-income populations."[7]

So let's make it available. The idea of requiring monitoring as part of the project permitting process would help address the fact that available statistical information may miss hot spots and many impacts, as well as the problem of having to reach a threshold of "substantial" before triggering action.

TRANSFORMING EJ NEIGHBORHOODS INTO GREEN COMMUNITIES

My student's interest in environmental justice remains strong, even though many of them are white and all are privileged to some extent, having found their way to a place in an elite institution. In 2018, we pulled together our thinking in a report about how to transform EJ neighborhoods into green communities. We figured that, by keeping dirty industries out (based on better monitoring) and attracting clean industries, planting trees, making parks, putting solar on roofs, insulating homes, and greening public transportation, all feasible and not requiring enormous budget-busting investments, people's lives could be radically improved.

Then one day a student[8] said that she'd been thinking about how the subsequent improvement in property values would drive up rents and the very people we wanted to benefit would have to leave the area. She said this phenomenon had a name: eco-gentrification. As a result of her insight, we transformed our project, and began investigating ways to boost job training and education, stabilize rents, foster small business development, and give the communities more of a say in what development occurred, including paying for work by residents—such as young people—that would be part of the community greening, so that they could earn enough money to remain in the neighborhood after it was improved. These ideas included community land trusts, paid community interest representatives, and a student-invented proposal for "a program that allows for companies to fund potential employees' education before they begin their higher education . . . an Income Share Agreement. During their final year of high school, students would apply to the program as well as universities in the field they would like to pursue in

7. Iliana Paul, Christine Pries, and Max Sarinsky, "Improving Environmental Justice Analysis," *Executive Order 12,898 and Climate Change*, January 28, 2021, https://policyintegrity.org/publications/detail/improving-environmental-justice-analysis.

8. Katherine Calandriello now working in community development.

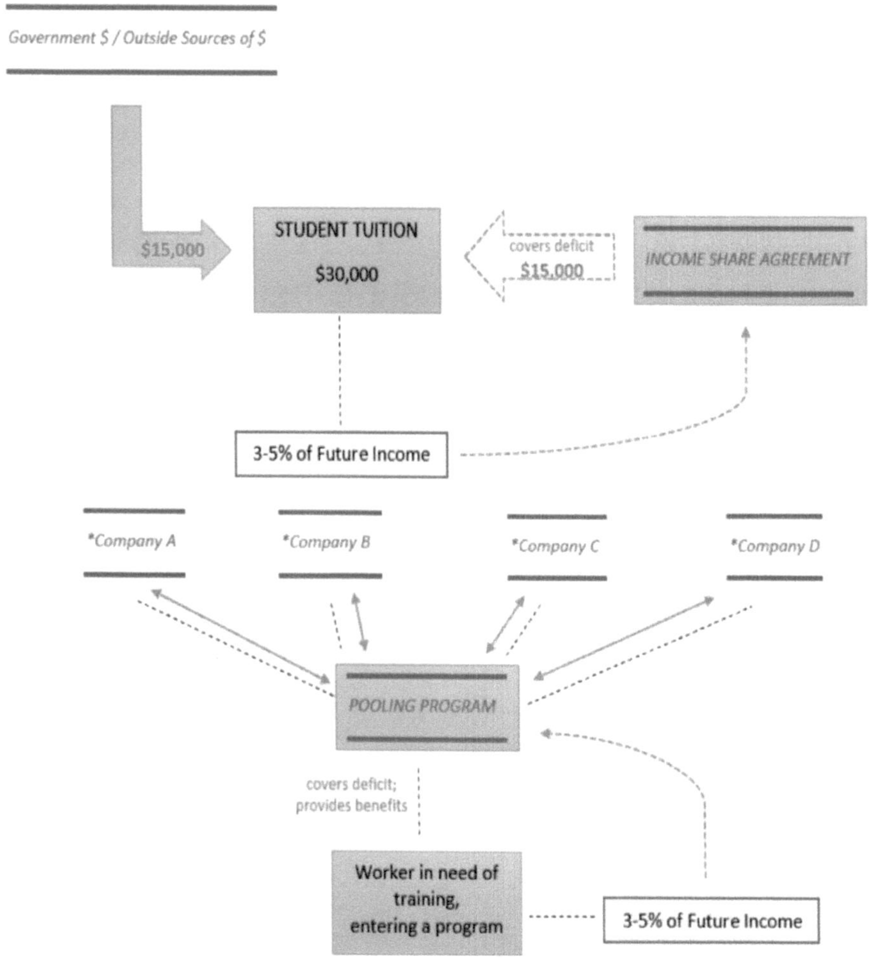

* indicates a tax break or government incentive given to a company for being part of a Training Sponsorship Pooling Program

Figure 8.1. The Income Share Agreement and How its Cost Can Be Distributed

the future."[9] Figure 8.1 is the explanatory graphic they produced to make their equitable cost-and-benefit-sharing proposal clear.

9. "Transforming Environmental Justice Neighborhoods into Green Communities," 29 and 30, www.bu.edu/rccp Fall 2018 Projects. The ideas about funding education were developed by Lydia Silber and Aseel Alharthi.

We presented these ideas to Boston City officials, who received them with much approbation—a tremendously affirming experience for students to have. After this one of the officials invited me to present at a meeting of the National Environmental Justice Advisory Council (NEJAC) meeting that was happening in Boston. At that meeting, I heard neighbors of vast chemical plants in Louisiana talk about the extreme conditions in which they lived, and the rates of cancer and other illnesses they saw and experienced. It was clear that during the Trump administration the deliberations of this council were not being heeded, but they continued anyway. At one point, the committee was asked, What does EPA stand for? Is it the Environmental Protection Agency, or the Environmental Pollution Agency? After I gave my testimony and walked past these victims, I could not find words to express the shame I felt for my society's insults to them, and my anger and helplessness. I regret that I did not simply stop and tell them that I was sorry, or thank them for sharing their stories, or any other attempt, however feeble or inadequate, at human connection with these people who have been so disrespected. It is easy to miss moments of opportunity to connect.

"OPPORTUNITY ZONES" OR COMMUNITY-LED DEVELOPMENT

The contact with NEJAC led to a meeting between my students and EPA's NEJAC coordinator, Matthew Tejada, who told us to look into "opportunity zones." My students[10] found that the city and state had designated Boston Harbor Islands and other places where people don't live as opportunity zones. This was bizarre! Why would the state do that? Our eventual conjecture was that they didn't believe the zones, as implemented by the Trump administration, would be of benefit to the poor. And the more we looked into the issue, the more this seemed to be the case. The zones had morphed from the original concept, proposed by Cory Booker, to something that would likely just benefit big developers. Communities weren't given a say in what would be built. Developers would be incentivized to build whatever they liked—casinos, for example. The students wrote to Boston City officials, suggesting a different approach:

> Our research leads us to believe that the best kind of development is that which is community-driven and considers the perspectives of the existing residents in the community where development is to occur.[11]

10. Christopher Reyes, Cory Seremetis, and Aseel Alharthi.
11. Community-driven Environmental Redevelopment and the Flow of Tax-Incentivized Funding, Spring 2019 projects, www.bu.edu/rccp.

Students continued to work this vein of ideas, one writing "Empowering Community-Based Development,"[12] which described several tools that can be used to meet the goals described, such as community benefits agreements between communities and developers, and guaranteed legal representation for those facing eviction or landlord retaliation, because it has been estimated that 90 percent of landlords have legal representation and only 10 percent of tenants do.[13] Students looked at community land trusts, a proven but underutilized model for helping residents stay in place and to build wealth, and share in the American dream not otherwise open to those just scraping by.

COMMUNITY LAND

The community land trust idea was a focus in a recent conference put on by the Urban Farming Institute (UFI) in concert with the Massachusetts Department of Agricultural Resources.[14] It is one way to create spaces for urban agriculture and ensure they are developed for the benefit of the community. Urban agriculture, another popular focus for my students,[15] is precarious, vulnerable to development and tax pressures. At the UFI conference presentations by Jillian Hishaw, who founded Family Agriculture Resource Management Services (FARMS), an organization that "helps Black farmers, and all farmers from historically disadvantaged groups in Southeastern states retain ownership of their land"[16] and Emmanuel Pratt of the Sweet Water Foundation, which practices urban regeneration in the South Side of Chicago,[17] raised consciousness of the sad history of displacement, first of Native American land, and then of Black farmers, and how we can and must reverse that history. Community land trusts have been used to foster a measure of home ownership for low-income populations. As described by Community-Wealth .org, a project of the Democracy Collaborative,

> Community land trusts are nonprofit, community-based organizations designed to ensure community stewardship of land. Community land trusts can be used for many types of development (including commercial and retail), but are primarily used to ensure long-term housing affordability. To do so, the trust

12. Summer 2019, written by Will Chang.

13. Russell Engler, "Connecting Self-Representation to Civil Gideon: What Existing Data Reveal About When Counsel is Most Needed," *Fordham Urban Law Journal* 37, no. 1 (2010), https://ir.lawnet.fordham.edu/cgi/viewcontent.cgi?article=2321&context=ulj.

14. See https://whova.com/portal/webapp/ufiac1_202103/ or contact Rose Arruda of MDAR or Pat Spence of the Urban Farming Institute.

15. See www.bu.edu/rccp, Fall 2019 projects.

16. https://civileats.com/2018/05/04/jillian-hishaw-wants-to-help-black-farmers-stay-on-their-land/.

17. https://www.macfound.org/fellows/class-of-2019/emmanuel-pratt.

> acquires land and maintains ownership of it permanently. With prospective homeowners, it enters into a long-term, renewable lease instead of a traditional sale. When the homeowner sells, the family earns only a portion of the increased property value. The remainder is kept by the trust, preserving the affordability for future low- to moderate-income families.[18]

Community land trusts can also be helpful in founding efforts that involve producing local food, providing healthy nutrition sorely lacking in many inner-city neighborhoods, often dubbed "food deserts," and to provide an amenity for those who live in often bleak conditions, as well as to provide many other community benefits, such as meeting places, and opportunities for practicing and learning crafts as well as farming.

Less than three hundred community land trusts are estimated to exist. But foreclosure rates are 90 percent lower than conventional mortgages. This is just one tool that can be used to help people in hard-hit areas, but it illustrates how many problems can be addressed at once, how we can turn EJ neighborhoods into green communities without displacing residents, and making sure that they get the benefit of the development that takes place.

ENABLING COOPERATIVE MANAGEMENT

Georgetown Law's Sheila Foster wrote in 2011[19] how we often approach the problem of degraded urban spaces by advocating stronger government intervention or privatization, but that both of these approaches carry significant costs, so nothing is done. However, the approach of cooperative development has worked. (Such as the Sweet Water organization, or the Eastie Farm in East Boston described by Natalia Tanko in "Connecting Farms and Food Justice Communities."[20]) Foster notes Elinor Ostrom's work demonstrating the success of "self-organized, cooperative natural resource management regimes"[21] and recommends government's role to *enable* those players who can organize collective action for the benefit of the community. Business improvement districts and park conservancies are examples. Her vision is one in which government must continuously evolve its role, gauging involvement to the capacities of the community, ensuring distributional equity and making sure things don't get frozen into place (ossification).

---

18. https://community-wealth.org/strategies/panel/clts/index.html.
19. "Collective Action and the Urban Commons," *Notre Dame Law Review* 87 (2013), https://scholarship.law.nd.edu/cgi/viewcontent.cgi?article=1001&context=ndlr.
20. www.bu.edu/rccp, Fall 2019 projects.
21. *Governing the Commons*, 1990.

Actively funding this function of government, and assistance to community members to be part of the development that benefits people, is direct investment in people, businesses, and community organization. The American Rescue Plan that President Biden signed contains an estimated $4 billion to provide direct payments of up to 120 percent of a socially disadvantaged, for example, Black, Hispanic, Native American, or Asian American farmers' or ranchers' outstanding debt, provide farm loans that historically have been denied to such farmers, pay off taxes associated with debt, and spend $1.01 billion to provide outreach, training, education, technical assistance, grants and loans, and funding to educational institutions to help improve land access for socially disadvantaged farmers and ranchers and address heirs' property issues, among other issues. Unfortunately, implementation as envisioned has not occurred.[22] Hishaw noted at the Urban Farming conference that in the early 1900s, Black farmers owned up to 16 million acres, and the 2017 USDA census estimates they now own less than 4.5. The losses in well-being and opportunity, shared by generations, are incalculable and are not made up by these provisions, however well-intentioned they are. But they are a step in the right direction, and it is important to know that every single Republican senator, except for one who was not present to vote, voted to strip these provisions out of the bill. I spend a lot of time trying to teach my students not to overgeneralize—about corporations (some act responsibly), about groups, about ideas, about people, and, especially, about Republicans and Democrats. But while realities are always complex, distinctions are crucially important. It is important to see that one party took this action, and the other opposed it.

WHAT ABOUT DIRTY FACILITIES?

Another question concerning environmental justice always comes up with my students, and that is: What happens when we tell companies they can't locate in an already overburdened area? This is a puzzle worth wrestling with. The simplistic answer is to say, well, make that dirty factory locate in a rich people's town. This, of course, violates the principle of universal respect, and raises the question of why it is fair to poison a child who, through no fault of his or her own, was born to a well-off family. But just because we balk at poisoning rich kids doesn't mean we can let that dirty factory operate where it hurts poor kids. We seem to hit a brick wall, but *that is only because we*

22. "A New Law Promised Debt Relief for Black Farmers. Instead, Some Got Collection Notices." April Simpson, March 3, 2022, https://inthesetimes.com/article/usda-farmers-of-color-debt-relief-collection-notices

have taken as a given the idea that the dirty factory should be allowed to operate somewhere as it is. We come back to the original compromise of our environmental legal system, which begins with a "permit" system. It begins with the idea that pollution is necessary, and we will permit it to happen, with a limit. We must go to the source of the problem to truly solve it. Pollution is often not necessary. Maybe it should not be permitted.

Does this solve the EJ problem? No, not right away, because we have immediate impacts and this is an evolutionary process. But if we prohibit any more dirty businesses in overburdened areas, institute far more comprehensive monitoring, and enable people to have more power over what happens in their neighborhoods, we will unleash forces that will accelerate the evolutionary change that we need to happen. The topic of environmental justice has been discussed for decades, but we have not yet made much of a dent in the problem. While prohibiting adding to existing burden is not a final answer by any means, it is part of a program of making our system evolve. A policy that discourages the growth of the dirty economy can also encourage the growth of a clean economy.

DISCRIMINATION BEFORE THE LAW

In "Vulnerability, Equality, and Environmental Justice: The Potential and Limits of Law,"[23] Foster explains the role zoning has played in causing polluting facilities to be located in minority communities, and that existing anti-discrimination and equal protection law is inadequate to the task of addressing this inequity:

> for claims brought under the equal protection clause, federal courts regularly dismiss discrimination claims for failure to prove explicit bias, or discriminatory intent, and failure to show a specific causal link between that bias and alleged discriminatory treatment.[24]

And even though the 1964 Civil Rights Act addresses practices that are discriminatory in effect, without requiring explicit bias,

> disparate impact claims have been limited by the requirement that a challenger must prove a specific causal link between a discrete policy of the decision

23. In Ryan Holified, Jayajit Chakraborty, and Gordon Walker, *Handbook of Environmental Justice* (New York: Routledge, 2017).
24. Ibid., 4.

> maker (an employer, public official, agency, etc.) and the statistical disparity that harms a minority group.[25]

Today's courts do not presume that bias is rooted in the system that is so obviously unfair, nor that unfair conditions are necessarily the result of existing policies, without specific proof. Making things worse is the tendency to limit review to whether there is compliance with the narrowest interpretation of the law. As discussed above, emissions are measured on a wide, averaged basis, and can miss that there are hot spots where particular people receive the worst impacts. Nor are cumulative impacts taken into account. But if the facility is in compliance, a court has to overturn an entire regulatory system to do justice. This jurisprudence makes it hard to bring about environmental justice. Foster recommends attention to the vulnerability of communities, referring to the arguments of Martha Albertson Fineman for a government responsive to people's needs, and utilizing information about the actual conditions that create the need for action. She notes that while EPA has developed a more contextual approach that would make room for using measures such as the EJ Screen or the Cumulative Environmental Vulnerability Assessment, its appeals process has not yet ruled in a manner that would enforce this perspective.[26] She notes, too, that measures that would create regulatory evaluation to bring about environmental justice would exist in tension with established measures for regulatory evaluation such as cost-benefit analysis and risk assessment.

SANITATION AS A BASIC RIGHT

Looking at environmental justice brings up so many issues: lead paint in dilapidated housing; high rates of asthma from cockroach droppings; high rates of pesticide poisoning from trying to kill the cockroaches; nature-deprivation; poor transportation; bad food; proximity to highways, factories, and contaminated lands; inadequate services; noise, and something that has recently gained deserved attention—poor sanitation. This brief discussion cannot do justice to the range or importance of all of the relevant issues, but ends with a mention of poor sanitation because it illustrates how the environmental injustice is not just an urban issue. In "America's Dirty Secret: The Human Right to Sanitation in Alabama's Black Belt," Inga Winkler

25. Ibid., 5.
26. EPA has stepped up efforts to use the EJ Screen to target heavily impacted areas.

and Catherine Coleman Flowers[27] explain how in Lowndes County, home to small rural communities that are largely poor and predominantly African American, what most of us take for granted does not exist for many. Here,

> Only a small share of the population is connected to sewerage. The majority of residents are supposed to rely on on-site sanitation systems, but soil conditions make these expensive to install and unaffordable to many households. Where systems are in place, they are often inadequate or improperly maintained, resulting in systems that back up, overflow, and expose residents to raw sewage.

Increased flooding due to climate change is expected to worsen this situation. Alabama state law simply says that property owners must install sewage treatment systems but it doesn't provide aid to ensure that this is feasible to do. Where municipalities have installed them, they are often inadequate. In a part of the country where the median income was below $26,000 as of 2015, and a system might cost $30,000 to install, a philosophy of unaided individual responsibility is unrealistic. It is "estimated that 25 percent of 850,000 septic systems in Alabama are currently failing" and 18 percent of households in the region have no system at all.[28] People may even be punished for not having proper sanitation, when they desperately want it but simply can't afford it. The population "has experienced a resurgence of parasitic diseases, such as hookworm, that are linked to poor sanitation," and a recent study found "that gastrointestinal parasites are present in >30% of this at-risk population in Lowndes County, Alabama."

In 2010, the UN General Assembly recognized a human right to sanitation.[29] In examining how the US sees this right, the authors recall FDR's "Four Freedoms Speech," in which he stated that

> "We cannot be content, no matter how high that general standard of living may be, if some fraction of our people-whether it be one-third or one-fifth or one-tenth-is ill-fed, ill-clothed, ill-housed, and insecure." The "second Bill of Rights" he proposed, however, never materialized, and socio-economic rights remain marginalized in the United States for a variety of political and economic reasons.[30]

---

27. *Columbia Human Rights Law Review*, http://hrlr.law.columbia.edu/files/2018/01/IngaTWinklerCatherineCole.pdf. Winkler is the former Legal Advisor to the United Nations Special Rapporteur on the human rights to water and sanitation and a lecturer in human rights at the Institute for the Study of Human Rights at Columbia University, and Flowers is founder and Executive Director of the Alabama Center for Rural Enterprise, the Rural Development Manager at the Equal Justice Initiative, and a Practitioner in Residence at the Franklin Humanities Institute at Duke University.
28. Ibid., 189–90.
29. G.A. Res. 64/292, 1 (Aug. 3, 2010).
30. Winkler and Coleman Flowers, "America's Dirty Secret," 201.

This is borne out in the actions of the United States in the UN, which consisted of expressing "a desire to join the international community in recognizing the human rights to water and sanitation," but not allowing itself to be legally bound by this recognition.[31] However, the US is a party to the International Commission on the Elimination of Racial Discrimination (ICERD),[32] and all parties to that agreement have committed to

> undertake to pursue by all appropriate means and without delay a policy of eliminating racial discrimination.

This international agreement recognizes disparate impact and social, economic and cultural conditions, addressed under the International Covenant on Economic, Social and Cultural Rights (to which the US is not a party, but which has entered into force). But while the US avoids creating rights that citizens can use to force government action in court, it can recognize the public interest in a healthy population not beset with water-borne disease resulting from the failure to control wastes that are demonstrably feasible to control. The authors state:

> In order to protect people from contamination, states must adopt and enforce policies and regulations, including those related to the use and management of septic tanks. However, states must not impose responsibilities that homeowners are unable to meet. Therefore, in addition to regulation, states must create environments that enable individuals to comply with regulations, taking into account the capacity, opportunities, and resources that homeowners have at their disposal.[33]

Because we have not acted, the United Nation's Special Rapporteur on Safe Drinking Water and Sanitation traveled to Lowndes County and reported to the world that

> More concerted efforts are required to ensure targeting of policies and programmes to reach the hidden and poorest segments of the population.[34]

The US is often referred to as the richest country in the world. We were the prime mover behind the formation of the UN. Our Eleanor Roosevelt led the creation of the Universal Declaration on Human Rights. It is not partisan or disloyal but an authentic response to bemoan how far we have fallen from

31. Ibid., 203.
32. International Convention on the Elimination of All Forms of Racial Discrimination, opened for signature, March 7, 1966, art. 2(1), 660 U.N.T.S. 195, 48 (entered into force January 4, 1969).
33. Ibid., 213.
34. Ibid., 221.

that position of moral leadership. Should we discover shame, and resolve to use the tools of government to recognize and give force to the basic rights of citizens, we should recognize that doing so will be good for everyone. It will reduce disease, and foster pride, in individuals and in ourselves as a society. The results of our cruel and foolish avoidance of responsibility cannot be calculated in any cost-benefit analysis. We can blame conservative, tightfisted, or racist politicians who turn their backs on those who are suffering and need help. But that would be only part of the story. If the rest of us don't care to act that is also part of the problem. Blaming others may only create denial, or worse, a backlash of resentment. The solution is before us: adequate funding from the federal government—assistance in meeting standards that apply to all citizens because of basic rights.

INDIGENOUS PEOPLES

Any reader apt to feel a sense of shame concerning how minorities, especially people of color, have been treated in predominantly white, Christian societies may wish not to be confronted with the issue of how Europeans and other modern societies have treated the more ancient groups they encountered in their spread throughout the world. Patriotic believers in the American system of democracy may not wish to acknowledge the abuses, the massacres, the perfidy, the cruelty in the long history of settlers and Native American nations, but it not only exists as an ineradicable stain on our honor it also persists. Until the 1924 Indian Citizenship Act, Native Americans weren't even considered citizens, and many still were denied the rights of citizenship in many states until the 1960s, and with the recent weakening of the Civil Rights Act of 1964,[35] the tools to combat voter suppression have been limited. I paid attention to *TVA v. Hill*, the 1978 Supreme Court decision that upheld the Endangered Species Act, protecting the little fish called the snail darter, but it wasn't until many years later, that I learned that the Tellico Dam that threatened the fish also flooded ancestral lands of the Cherokee, from which they had been driven, to walk the Trail of Tears, thanks to President Andrew Jackson. While we do remember that Chief Justice Marshall authored an opinion in *Worcester v. Georgia*[36] that stated that Native Americans retained their natural rights, rendering Jackson's actions unconstitutional, principles did not win the day, however, and it is incumbent upon those who wish to see a better world to recognize the failures of the past and to ask both, How was

35. *Shelby County v. Holder*, 70 US 529 (2013).
36. 31 US (6 Pet.) 515 (1832).

that crime allowed to happen?[37] and, How can we prevent it in the future? It is surely part of the story that America was seen as a land of opportunity for the new arrivals, who craved the land to grow such things as cotton and sell it for money. But it is surely another part of the story to recognize that Native Americans were considered by too many to be savages, and that our history is one in which another people was treated as if it simply did not, or should not, exist, and that taking of their land was righteous.

This overdue recognition is still necessary because as we take a look around the world today, we find we may still be living in a way yet influenced by this primitive past. The town of Kivalina, Alaska, sued the major oil companies for causing the climate change that was causing the rising seas making it impossible to continue to live on the coast. The case was dismissed in 2009, because the court felt it didn't have the power to act, that it was up to Congress and the president.[38] But what about when the Congress and the president don't act, as they had not and have not yet done, to address the need to relocate the residents of this town? Maybe the evidence of failure to act to relocate residents in other coastal areas threatened by sea level rise is evidence that the failure to act is not essentially racist, and millions have been provided to relocate the Biloxi-Chitimacha-Choctaw, a Native American tribe living in the Louisiana coastal wetlands, who have lost 98 percent of their land.[39] The people of Kivalina have been engaged with many groups—agencies and nonprofits[40]—but the planning of an "equitable retreat" from rising seas is largely absent, so that at this point an observer can comment: "Retreat is in fact happening to the poor; we're seeing injustice being reinforced through retreats. New properties are being built for the wealthy along the coast, while the poor are being moved away from it."[41]

This impression that certain peoples are not priorities is reified by a recent article in the *New York Times* titled "A Global Plan to Conserve Nature is Lacking a Vital Ingredient." The subtitle of the article spells out what that missing ingredient is: "Talks to protect 30% of the planet by 2030 won't include indigenous people." Negotiations are planned for later this year in which dozens of nations will hammer out a plan to preserve old-growth

---

37. See, for example, the 2016 *Jacksonland: President Andrew Jackson, Cherokee Chief John Ross, and A Great American Land Grab*, by Steve Inskeep.

38. *Native Village of Kivalina v. ExxonMobil Corporation et al.*, 4:08-cv-01138, September 30, 2009, US District Court for the Northern District of California.

39. https://news.yahoo.com/native-american-tribe-relocate-louisiana-coast-sea-levels-231726681.html.

40. https://toolkit.climate.gov/case-studies/relocating-kivalina

41. Jola Ajibade, interviewed by Katherine Bagley, in "Equitable Retreat: The Need for Fairness in Relocating Coastal Communities," *Yale Environment 360*, November 19, 2020, https://e360.yale.edu/features/equitable-retreat-the-need-for-fairness-in-relocating-coastal-communities.

forests and wetlands that store carbon and filter water: an excellent idea. But without including the people who live on that land, what can we expect?

George Catlin was a young lawyer who tired of paperwork and headed out West to live with Native Americans. Thanks to him, we have records and paintings of various tribes he visited.[42] In 1841, Catlin proposed a "Nations Park" that would not only preserve the wild and beautiful prairies full of buffalo but also the way of life that he witnessed there—the peoples themselves. That it wasn't done then doesn't mean there aren't opportunities to do conservation that way now.

TWO FORMS OF KNOWLEDGE, TWO FORMS OF WASTE

What is the most important difference between Indigenous Peoples and modern peoples? It is not that developed societies are smarter or more sophisticated. They have created modern technology, but more ancient societies have what biologist and Potawatomi Nation member Robin Kimmerer calls TEK—traditional ecological knowledge.[43] We need to learn as much from them as we presume that they need to learn from us—more, in fact, at this point in the turning of human history. They have retained connection to the land. We have lost it. This is not knowledge we can take from them, or patent and sell. It is wisdom we may acquire, beginning by respecting them.

In *Changes to the Land*, William Cronon explains that the European arrivals felt that the Native Americans were wasting the land by not developing it. (This was a concept articulated by many, including John Locke). The Native tribes exercised what the Europeans regarded as property rights over their agricultural plots, and the rest was owned by no one, but open for all to travel through. In contrast, the Europeans fenced off land, exercised a right to exclude others from what they considered they owned, and treated the land as a commodity to be used to produce things to sell. They let their pigs roam freely, however, and they ate the Native plantings not protected by fences.

PROTECTING LAND BY HELPING FIRST PEOPLES TO REMAIN ON IT

After being sent by EPA in 1998 to Brazil to train the environmental agency of São Paulo in pollution prevention, I became friends with Paulo Coutinho,

42. See https://www.georgecatlin.org/the-complete-works.html.
43. See Robin Kimmerer, https://www.esf.edu/faculty/kimmerer/;and National Park Service https://www.nps.gov/subjects/tek/tek-vs-western-science.htm.

the official of the agency (CETESB) in charge of the training. Paulo and I have continued to correspond since that time. Around 2016, he sent me information about a project he was working on with an Amazon tribe, the Yanomami. The project was for an agroecology initiative, designed by chief Dawi Kopenawi, to keep the young people of the tribe on the land, so that they would not migrate to the cities, and the tribe decline. Coutinho said that the land was being invaded by "bandits," artisanal miners, and others, some violent, clearing the forest, damaging the land, and that the chief felt that if he could keep his people on the land, they could protect it.

In the class that I teach called Developing Sustainable Environmental Responsibility,[44] I asked my students to consider what responsibility they might have toward this far away people. I shared the information about the agroecology project and asked, Is there a way to help them? Is this something we ought to try to do? To help keep Indigenous Peoples on their land? It seemed to me that this was something that benefited us all: that the Amazon is something the world needs to remain intact. And that here was a people willing to fight to keep it the way it is. That there is a moral reason to help them, as one can understand by asking how you would like it if it happened to you, if you were being pushed off your land? And that there is a practical reason to help them, that we all need them to stay on their land so that it can be protected as a carbon sink, as a reservoir for diversity, and as a preservation of unique human culture that we ought to see as valuable, special, precious. They can resist the incursions of artisanal miners and clear-cutting cattle raisers and soybean planters. It seems to me that we can help them do that but didn't state these concepts until first asking the question to see if students felt that way.

Several of the students did feel a sense of responsibility and a desire to help, but not all felt they had the capacity to do anything. Some were too busy struggling with how they could reduce their own footprint, and how they could get conservative voters in this country to stop denying the reality of climate change. For them, the question was too overwhelming, the problem too distant. But some did wish to grapple with the problem. As it turns out, it is not so easy to figure out what to do.

I wrote to Paulo saying there was a student willing to try to raise money through the Brazilian Student Association, which exists on many campuses. But there was no way to transfer that money. It took a long time to get information about a bank account. But that only raised questions about control of that money, and I couldn't endorse sending money to someplace with

44. The concept is: What responsibility will we accept and expect of others, that we can sustain, and which, if observed, will help us attain sustainable societies?

no assurances about how it would be handled. Each question led to more questions.

I raised this issue in successive semesters and with the students in my project class as well, and some engaged in the effort, contacting people who work in the field (Mark Camp of Cultural Survival, in Boston, was of special help). In 2019, we produced as a work in progress a draft *Guidebook for Donations to Indigenous Peoples of the Amazon*, but when I asked for reviews by experts, I was told by one that the very concept sounded paternalistic, reflecting a privileged white perspective, that assumed a one-way flow from us to them. I felt this failed to recognize the intentions of people who simply want to help, and their good reasons for doing so, but I took the comments to heart and rewrote what the students produced. The next iteration was a report that bore a question as its title: "Can Citizens of Industrialized Nations Help Indigenous Peoples Keep Their Land?"[45] In December 2020, the organization Resilience published a piece I wrote describing the report.[46] I received an email that read:

> Dear Rick,
>
> My name is David Good. I am a Yanomami from the Upper Orinoco region of the Venezuelan Amazon, member of the Irokai-teri community (above the Guaharibo rapids). I was raised and educated in the United States and coordinate a US based nonprofit that works with the Yanomami communities in the Venezuelan Amazon. I just read the guidebook "Can Citizens of Industrialized Nations Help Indigenous Peoples Keep Their Land?" It is a wonderful guidebook and thank you very much for putting it together. I am a lecturer as well (biologist), and each session ends with my students with "Well, what can I do all the way from here?"
>
> Would you be willing and available to connect by phone or Zoom for a virtual meet and greet? I would like to share the work of the Good Project and my research. We are a small, grassroots, nonprofit that supports Yanomami programs in the Upper Orinoco of Venezuela. These communities receive very little international attention and support. I'm well aware that the political and economic crisis in Venezuela is a major factor. However, we are a highly specialized team with decades of experience working with the Yanomami and the Venezuelan government and have been able to execute very important health projects and research in the last couple of years. On a personal level, I share my life experiences living, working, studying with the Yanomami people, my Yanomami family, to help raise awareness on the importance of protecting their culture, lands, and rights. Storytelling is my passion.

---

45. www.bu.edu/rccp, Spring 2020 projects.

46. https://www.resilience.org/stories/2020-12-10/helping-indigenous-peoples-stay-on-their-land/.

When we talked (over Zoom), it was clear that David was just like anyone else who grew up in New Jersey—an apparently fully Americanized, modern person. But he is also Yanomami. He told me that spending time with his mother and his family there was wonderful, that life was peaceful and relaxed, nothing like how we live here. But they live under many threats. Check out these links:

www.jointhegoodproject.org/expedition2022
www.jointhegoodproject.org
www.davidgoodamazon.com

There are ways to help, to engage, to learn from each other. The arrogance of modern civilization is an obstacle, but why should we regard it as insurmountable? The ancient peoples have found their ways of life worth preserving for a very long time. There is something in that, something we must learn.

Concerning the system built around making money, capitalism, many have observed that it is unsatisfiable. Money expands infinitely. You can never get enough, and we should observe billionaires not as people who are succeeding but people who are failing, because they act as if they have some unfillable hole in their souls that they cannot fill. Why they don't say "enough," and take pleasure in sharing—without conditions—what they never had a right to, is a puzzle of human nature. But human nature is not limited to such characteristics. Other societies show us that. They demonstrate the possibility of sufficiency, of satisfaction. This may be more important to our fate than the concrete fact that many have also shown how to live in harmony with their surroundings, regarding other living things as relatives, and seeing a responsibility to maintain a balance with them. We, the supposedly advanced society, are the ones who appear as children with these perspectives in mind.

Law professor Tony Arnold has written:

> Our most pressing environmental problems are fundamentally moral, spiritual, and social issues, involving an unjust order shaped by structural inequality, systemic racism, and oppression. The environmental conservation movement has too often tended to believe that our salvation is rooted in science done by white men at elite universities in the Global North and too often given only superficial attention to voices and contributions from the Global South, marginalized peoples, and non-scientists.[47]

47. Posted on a LISTSERV, shared with permission.

CONNECTION

I once was able to obtain a small grant from EPA, back in the day when that agency had money to give out, to work with the tribes of New England on their waste issues. I had some ideas about how waste generation could be reduced and how existing wastes could be recovered and dumps cleaned up. I had the idea that the land grant colleges could hire them to teach their concepts and knowledge of nature. (There was interest from the University of Maine but the grant was tiny and I was not able to pursue this beyond demonstrating the idea's potential.) When I visited the Penobscots on their reservation in Old Town, Maine, I sat in a room and waited with some tribal members for others to arrive. One of them gestured, taking in the room in which we sat, and said, "we've been meeting right here for thousands of years." Thousands of years. I thought of how my knowledge of my own family goes back only two generations, and how when I would ask my grandmother about my ancestors, she would say, "that was the Old Country," and had no desire to say any more than that. This was, of course, because the Jews were chased out of there, and were lucky to escape what came shortly after.

I thought of this with some pain at that moment, but after our meeting, I was invited to attend a ceremony of dedication of a new building for community events. I stood nearby while an elder asked everyone to gather around. And then he began chanting, and it was exactly the same experience as being in a Jewish temple, with the rabbi chanting an ancient language before breaking the bread and sharing it. The food was shared with everyone, the same ritual. This was a very different culture, far away from where the Hebrews developed the traditions handed down to me, with striking similarity. I felt an uncanny sense of connection. I believe this feeling is open to anyone who seeks engagement with those they might think of as "the other." I can weep to think of how often this opportunity has been missed, but I can also exult in the knowledge of the chance for understanding the unity that can bestow healing.

Rebuilding environmental governance must be about establishing relationships, especially where they have been broken.

THOUGHT EXPERIMENT

Imagine if modern intruders into the lands of First Peoples—those in residence for generations—had learned to coexist. What would that world look like? Might we still have the passenger pigeon, the mighty herds of buffalo? The great towering evergreens of the East? Flourishing wildlife, a rich ecology, beauty?

Now imagine how the wisdom of how those peoples coexisted with nature can change our modern ways of life. Imagine how we might interact with First Peoples with respect in order to bring that about. Imagine modern people wanting to truly learn from the ancient peoples. Then compare that thought with the attitude that they should only learn from us, they should become like us.

Chapter Nine

The Fixer-Upper

When you step back and take a look at the house of environmental protection, it can look like an old house with peeling paint, leaking gutters, and drafty windows. It's dilapidated and needs work, but it's basically sound and a great bargain. It's a fixer-upper. It's not a teardown. The foundation has not shifted, the walls are only a little cracked.

The chapters up till now have touched on very large issues and principles. Small issues can illustrate choke points, illuminate obstacles we can remove, and show us how to become a more rational system. When the system doesn't make sense to people, when rules make people shake their heads in disbelief with good cause, that's like rain leaking through the roof we thought was there to protect us—the system under which we operate. A system seen as rational would have a better chance of inspiring faith and willing compliance. Maybe we can see the wholesale changes we need but do not yet have the political will to make them. At least we can fix up the house we have, to keep the rain out.

THE SUCCESS OF LITTLE RIO

Inspired by Carol Browner's Common Sense Initiative, I proposed a Regulatory Improvement Opportunities (not long after the Rio Summit, so I called it Little RIO) event in Massachusetts, and after much foot dragging and questioning about why the agencies should expose themselves to criticism, was given permission to go ahead. I sent out invitations to the business community and interested parties about what bothered them and they might want to change. I got a few dozen immediate responses, and while some were angry about the issues they complained about, all were excited and happy to be asked. I talked about the issues with them beforehand, so we could make up a clear statement. Some people had to be helped to phrase their concerns in specific terms. One fellow I talked to took forever to make clear that what was making him yell and use expletives was the fact that he was required to report to two different authorities that he had backflow prevention on his water lines when it seemed to him that he ought to only have to report to one. (He was right.)

With this processing up front, to frame each issue as a clear question (Is there a duplicative requirement here that needs to be fixed?), we were able to have a constructive discussion and the business attendees left with high praise. I was thanked many times. The gratitude expressed by the regulated community for being asked what they think was huge.

But it was not very gratifying for the few officials in attendance. Only one said anything to me. She told me that it was hard, very hard, to take that look at the regulations and see that yes, there was a problem with them. There really was a palpable sense that the agency people needed to believe their rules work fine and any problems are misunderstandings by the public. It was hard for her to see how that was not the case. She said it was hard, but she was very glad that she came. I wrote up a report and shared it widely and received no other response, though I believe at least some of the issues were taken up and addressed. Though it received the highest positive response of any of the hundreds of events I'd done for the program, no one urged me to hold another Little RIO. The DEP official whose eyes were opened soon left government to work in the private sector, where she helped her company be more responsible, but the public system lost someone who might have helped the agenda for change. The encouragement for institutional reform has to come from the top down, which will be more likely to happen if the public demands it.

THE ZERO WASTEWATER DISCHARGE CONUNDRUM, OR, THE CLOSED LOOP CLOSED DOOR, AND THE NEED TO PRIORITIZE REGULATORY RATIONALIZATION

There was one unfortunate aspect of our environmental regulations that came up repeatedly over the decades I spent in government: the Closed Loop Closed Door. It caused some real consternation and, seemingly, could not be fixed. The fact of it symbolizes to me that inability to repair the leaking roof. It is not a problem that is causing great physical harm, but its existence makes the whole structure of environmental regulations seem less rational. If, when companies do something good, we have to tell them, "Oh no, the regs don't allow that," that does not send the message we want sent, if we want industry to keep evolving. Opening the closed door to something we want—closed-loop systems—should be a priority, not because there's an emergency, but because we should have a system that continuously improves, that fixes itself when it isn't making sense.

The closed-loop closed-door problem emerged from the first great success story of the Massachusetts source reduction program: the Robbins Company, a maker of awards and decorative items in Attleboro, which stunned everyone by achieving "zero discharge" of the wastewaters from its metal-plating operation, recovering metals instead of sending them out as pollution. Attleboro was one of the places where the metal-plating industry originated, and when, in 1987, the state gave notice that because of high levels of pollution in the Ten Mile River, it would have to tighten discharge limits, many companies were affected. Paul Clark, environmental manager at the company, told me at the meeting that when this was announced, one of the attendees said that the only way to meet the tightened restrictions was to stop discharging completely. Clark said that people laughed, because this was impossible, but that night he went home thinking, was it possible?

Clark had been hired not long before because the company had been in constant violation of the more lax requirements, and he had brought the company into compliance through standard treatment methods, such as collecting wastewater in large vats and adding chemicals to cause the metals to precipitate out into a sludge. Clark realized that it would cost several hundred thousand dollars to expand treatment to comply with the new requirements. At a number of workshops put on by the state's Office of Safe Waste Management,[1] Clark learned about using flow restrictors and countercurrent rinsing, techniques that reduce water throughput, and got his wastewater down to 13 percent of what it had been before. He also learned about metal recovery techniques, which now made a lot more sense because the water conservation efforts meant his metals were now much more highly concentrated. With the support of the company president, Robert Chatel, and consultants, he installed inline treatment that destroyed the cyanide used in plating, neutralized the acids and caustics, and removed particulates with filtration and organic chemicals with

1. OSWM was part of the Department of Environmental Management, which under Dukakis' Commissioner James Gutensohn focused on "source reduction" of hazardous waste, which was a result of years of planning and policy development. OSWM was created to test the concept of government assistance to help companies and others learn to prevent pollution—at that time it had a vigorous Household Hazardous Waste program as well. While OSWM was transformed into OTA, the policy and planning section of DEM no longer exists. It is this author's opinion that policy is more influenced today by the predilections of the administration, than before, when more ideas flowed upward from the middle-level staff (many at OSWM moved on to careers of highly respected achievement in the environmental or public health field).

carbon. With wastewater treated to this extent the company could now use ion exchange to capture metals. The ion-exchange system was then periodically regenerated, and the metals captured by them were now so concentrated they could be electrolytically plated out to be used again as pure product. An evaporator then boiled the remaining sludge down to an average of 7 gallons per year (from 40,000 gallons previously). Because the metals in the sludge were now so highly concentrated, it was valuable, instead of a waste.[2]

The system cost $220,000, which was much cheaper than what a conventional wastewater treatment system would have cost. It saved the Robbins Company $117,000 a year, just comparing to previous wastewater treatment costs.[3] But as Robert Chatel told me when I interviewed him for one of OTA's first case studies, this was nothing compared to the value of having pure water, which the treatment system now produced, and which he could now use to do his plating. Before, he said, the company used "city water"—what came in from the city's water supply. It was variable in quality and sometimes that disturbed the plating process. When quality control says something hasn't plated right, you have to strip off the surface and do it again. This generates a lot of toxic wastes and uses up valuable production time. But when you have treated your own water with all the filtration and carbon and ion exchange, it is of a constant, high quality, and your "reject rate" drops dramatically.

We were thrilled with the Robbins company story, and when I had Paul Clark speak at one of our events, he ended it by pulling out a copy of his wastewater discharge permit. It said there was *no discharge*. The local sewage treatment plant was not receiving a drop of industrial process wastewater from the company anymore. They still received wastewater from the bathrooms and a cooling system, but that was it.

We told the "zero wastewater discharge" story to anyone who would listen, and the story reached Governor Michael Dukakis, who was then running for president, and he paid a visit to the company. And then, because George H. W. Bush was competing with Dukakis on environmental grounds (Boston Harbor had not yet been cleaned up and Bush pointed to that objectionable water with good effect), Bob Chatel was invited down to the White House and honored there. Up until now, this is just a typical pollution prevention success story, interesting because it was one of the first, because it was so successful, and because people in power paid attention.

---

2. https://p2infohouse.org/ref/01/00165.pdf.
3. https://p2infohouse.org/ref/05/04211.htm.

THE HERO IS A CRIMINAL

But then the dark underside emerged. At around this time I was not just working to build trust in what OTA could do, in the approach of source reduction, but I was also working to build trust at DEP, the enforcement agency, in our new approach of working with companies. We were having regular meetings where I was helping them to understand that our assistance efforts were not upsetting the applecart. Many enforcement people were worried about what we—"the government," but another agency they could not control—were advising people. It was helping assuage concerns by meeting with them frequently so that we—the assistance staff—were very clear on the law, and so that they—the enforcement staff—had some sense of what was going on (within the limits of confidentiality requirements).

Robbins was telling its story, there was no confidentiality issue there, and I presented all the details, expecting shouts of praise and exultation. But one of the hazardous waste specialists pointed to the last stage in the process—the evaporator. He said, "That's treatment without a license."

I said, "What?" He explained: only licensed hazardous waste treatment facilities could do something like that. I said, "But there's nothing in that final water but metals. The chemicals have all been filtered out." He said, "It's still a hazardous waste,[4] and all that other treatment—it's also treatment without a license." In the old system, when there was a discharge to the sewage treatment facility, then all the treatment processes were in compliance with Clean Water Act treatment rules. *But once the company was no longer discharging to the sewage treatment plant, it was now under the Resource Conservation and Recovery Act (RCRA) laws, the hazardous waste law, and that had different rules regarding treatment, and the company was in violation of those rules.*

When I next met with the company president, who was just as pleased and proud of his story as he deserved to be, I had to tell him that technically, he was in violation of the rules. He was shocked and said something like "you mean I'm a criminal?" This was at the very same time that he was being honored by the two men running for president.

DEP was kind enough not to make an issue of this, because they knew it was wrong to do so. They saw that the example of Robbins was having a great effect, inspiring other companies to try to reduce their discharges. But the problem was real, and we convened a workgroup, calling in EPA as well, to figure out how to resolve this issue. We met for two years and accomplished nothing. Basically, neither the state nor the regional office of the EPA had the authority to suspend the rules, which were written in the federal legislation.

4. The "derived from" rule says that anything derived from a hazardous waste is a hazardous waste.

It was up to Congress to change the law. We wrote memos to headquarters and nothing happened. It was my first experience with the particular form of frustration that stems from real problems in the law. It's different from being frustrated because justifiable requirements are hard to meet. These were not justifiable requirements. There was something wrong with the law itself and we couldn't fix it. The treatment allowed under the Clean Water Act was different than that under RCRA. To change it we would have to change federal legislation.

INTEGRAL SYSTEMS

The DEP never went after Robbins as far as I know. I did learn one thing, though: if you didn't have any air emissions either, and you connected all the treatment steps, you would have a completely closed system, and that would be exempt from RCRA. Robbins's evaporator was "atmospheric," open, allowing water vapor to escape (and hopefully nothing else). So from then on, OTA recommended completely enclosed systems. We found companies that made enclosed evaporators (often termed "vacuum evaporation") and we recommended those. We made sure that if a company wanted to do something like what Robbins did, that it would tightly connect every part of the treatment system so it was all "integral." That made it like part of the production process and was how you could do in-house treatment under RCRA. This worked for some well-capitalized companies who had the time to research this.

But this was not always feasible for companies. We should want to get every industrial wastewater discharge into a closed-loop system, and if only water vapor is escaping, and that was good enough for the Clean Water Act treatment process, can it be good enough for the RCRA treatment rule? If companies stop discharging industrial wastes to the POTWs, where it mixes with everybody's personal wastes, we can put sludge back in the soil without such risk of contamination as we now have. Without industrial discharge the microorganisms that break down our wastes will function far better. Without industrial discharges we have fewer pollutants passing through the system to pollute receiving waters. We should have rules that encourage the evolution of zero discharge from industrial sources.

A TEMPORARY FIX IS NOT GOOD ENOUGH

About ten years after the Robbins company story, I got a call from John O'Hare of the Greater Lawrence Sanitary District about a company that had stopped discharging to his POTW because it had "closed the loop." When I visited the company, I saw open treatment, not integrated, and I told everybody about the RCRA requirements, and the company said it would only be water vapor and John said, don't worry, even though they are not discharging to the sewage treatment plant, he had issued them a "Zero Discharge Permit," so they were still under the regulatory rubric of the Clean Water Act.

This sounded like a great idea, but maybe something like a sham or trick, so I asked if we could go over this with the enforcement authorities. This then led to another round of meetings with DEP and EPA, the result of which was that they agreed to let John issue Zero Wastewater Discharge permits. All's well that ends well, I thought. We can live with this flaw in the system if the enforcers won't take action.

But I worried that there aren't many people like John O'Hare at POTWs, many of which don't have much staff, or people working on water who know about RCRA and vice versa, so when I was invited to be part of the Common Sense Initiative, an EPA effort to look at rules that ought to be fixed, I brought it up. EPA's administrator, Carol Browner, had asked invitees to think about how to make EPA work "Cleaner, Cheaper and Smarter," and I thought that if we could come up with a good resolution of the Closed Loop Closed Door that this would promote zero wastewater discharge, and would also make the EPA look much more reasonable. When a Bob Chatel had to wonder if he was a criminal or a hero, that was not good for EPA and its mission.

My suggestion was taken up and the Common Sense Initiative worked on this problem for a few years. We hired consultants to do studies to verify that what I said was true. They did. We brought in experts from the Clean Water Act and the RCRA programs. They confirmed the problem. No one came up with an answer. I was honored to be selected to make presentations to the Common Sense Initiative Council, made up of agency chiefs and heads of major environmental groups and other major figures. I laid out the issue, reporting that it remained unresolved. I was subjected to questioning that to me expressed skepticism and even a sense of annoyance by the chiefs of the water and hazardous waste departments.

Having failed to resolve the issue now in these three iterations, it receded, as it didn't seem that anyone wanted to go after the companies who everyone could see were doing a good thing. But then, about a dozen years later (2012 or so) I got a call from the company that John O'Hare had issued his first Zero Wastewater Discharge permit to. They had been hit with a huge fine from EPA.

I called up the EPA attorney and asked for a meeting. He didn't have to meet with me, but he did, and I told him the whole story. He was sympathetic but said his hands were tied. They were violating RCRA. I told him about how we had met with EPA about this very same company and EPA had approved the Zero Wastewater Discharge permit.

A few weeks later, he contacted me to say that no one could find any record of our agreement about John O'Hare's Zero WW Discharge Permits. No one remembered it. I went back to my office and hunted through files. I couldn't find anything that said EPA had approved the permit, but I did find communications that indicated maybe that was what they did. I had another meeting with him and with his boss, who by chance had been my supervisor when I worked at EPA for a few years. They listened and were convinced that maybe I had a point. But again, their hands were tied.

There is a certain amount of enforcement discretion that can be applied in such cases. One change, if I were in charge, would be to do more to investigate the motivation of companies found in violation. A company that is trying to do the right thing should get credit for that. This company was trying to be environmentally responsible. You couldn't just look at them and say, they're violators. Doing that would not just be unfair to them, but it would discourage others from even trying. We have to reward the effort and not just look at the outcomes, if we want progress. We ought to make big distinctions between technical violations and real bad acts.

Because these attorneys were good people, that's the approach they took. The company got off with a very small fine, and they had to enclose their process. This cost them some money but it's not a bad thing. It's safer, it prevents any untoward air emissions, and reduces the chance for spills. In an ideal world, companies would receive lots of financial assistance to put such things into place. This company was not happy they had to spend that money, and spend it then, instead of when it fit into their business plans. But they were happy they didn't have to pay a big fine.

STEPPING BACK TO STEP FORWARD

To my knowledge, the zero wastewater discharge conundrum has not been solved. RCRA says that if something is adequately covered by another law, such as the Clean Water Act, then it is exempt. But if a company is not discharging, it is covered by RCRA. And further: if you were once covered by a Clean Water Act permit and you stop discharging, there's an argument you can still be covered by the Clean Water Act. That's the logic of the John O'Hare Zero WW Discharge permit. That was the logic by which DEP refrained from prosecuting Robbins.

But what if you want to build a zero wastewater discharge system from scratch? You start operating without ever discharging. You never got a Clean Water Act discharge permit. These companies are not the same, and could be treated more strictly.

It seems to me that we can write rules that make sure these things are safe. We can look at treatment under the Clean Water Act and RCRA and harmonize them so this problem goes away. We could also revisit the iron rule of RCRA that all treatment be done by licensed Treatment, Storage and Disposal facilities. These are difficult and expensive treatment licenses to get so small entities don't seek them. Channeling treatment to only a few big places stems from the idea that we want all hazardous waste to go to places where treatment is done right. But we have opportunities to avoid shipping hazardous wastes if we maximize the opportunities for on-site treatment (made safe through regulation). This has happened for some simple treatments, such as neutralization, but not for closed loop systems, except, to my knowledge, in California.[5] It will also cost some money, but it would be worthwhile to provide financial and technical assistance so that companies will do it right.

Eventually, where we need to go is to systems that return the nutrients that we excrete to the soil. We are sources of compounds that feed plants that should be returned to the earth in an eternal cycle. While the building of sewage treatment plants was a boon for humanity, reducing waterborne diseases, bringing about one of the greatest improvements in human health in all history, it is an unsustainable system, that leads to overloading of our waters with nutrients, causing algae growth and eventually dead regions. Going with that overloading is an impoverishment of our soils. The system takes phosphorous, for example, in our urine, and sends it out to sea. But we need it to grow plants. This cannot continue.[6] We will not have an easy time closing the circle on our own discharges until we separate out the unnecessary contamination from industry.

Redesigning the technological systems that we now depend on is one aspect of this problem. But redesigning governance is fundamental. When a

---

5. California apparently found ways to do it https://p2infohouse.org/ref/05/04838.pdf;

California businesses using an ion exchange system to recycle its rinsewaters do not require a tiered permit if certain conditions within California's recycling law are met. However, evaporation of the regenerant will most likely require a tiered permit, either a Permit by Rule or a Conditional Exemption-Small Quantity Treatment permit depending on the quantity of regenerant evaporated." https://www.pfonline.com/articles/approaching-zero-discharge-in-surface-finishing.

6. "In 2017, the German government issued a new sewage sludge ordinance stating that from 2029, cities with 100,000 inhabitants or more must recover phosphorus from sewage. Cities with 50,000 or more residents have three extra years to comply," https://www.dw.com/en/phosphorus-in-our-pee-the-new-gold/a-53225198.

company can safely treat its wastes on-site, it should be allowed to, and we can design tailored regulations that ensure this is not abused. We do need more staff—to provide that help—and more inspections to ensure safe processing. We can also use government to provide help, to get more companies to go zero discharge. We can provide training, education, one-on-one technical assistance, research and demonstration, and financial help. We can make sure that a company applying for a "zero-d" permit gets their application expedited. We can make sure that permit fees are not a hindrance. We can provide recognition and purchasing preferences. We can create incentives for them to operate, and tighten up on competitors who continue to pollute, even though safer alternatives have been found and demonstrated. We can do a lot of things, but the one thing we should not do is continue as we have, leaving regulatory obstacles to zero-d in place, the very confusing nature of the rules an obstacle.

In this mismatch between how treatment is regulated under RCRA and under the Clean Water Act, we see the limits of the rules' original logic and the opportunity to reframe the problem to incentivize something we need to evolve. We see the possibility of structuring a unified approach rather than one divided by media. A rule about safe treatment could reduce both discharges and hazardous waste shipments.

IMPROVING COMPLIANCE RATES

The Clinton administration's Reinvention campaign focused on eliminating waste and improving regulatory governance, and when in 1993, Al Gore announced how this would be done, he "unveiled his initial recommendations for cutting red tape in front of two forklifts piled with stacks of regulations on the White House lawn."[7] It wasn't just that there are a lot of rules and you could never feel that you had covered all your bases, no matter how much investigation you did, it was also that they were difficult to understand. Take identifying your hazardous wastes as an example. You are supposed to look at a list, and see if the substances there are in your wastes. That takes a long time to do the first time you do it. If you aren't a chemist and don't have complete knowledge of the ingredients in the things you use (few do), you can never be quite sure of where you stand. You are also supposed to determine if your wastes have any of the characteristics of hazardous waste. If you do this right, you do tests to see if your wastes are corrosive, toxic, reactive, or ignitable. And what if your waste stream varies from day to day? The stakes are

7. https://www.nytimes.com/1994/12/19/us/gore-the-soldier-of-streamlining-returns-to-lead-clinton-s-charge.html.

high—ignorance is no excuse—and the expense of having sent your wastes to the wrong place could put you out of business. Now from the outside looking in, of course a business is responsible for the wastes they produce. But from the inside, people did not feel secure they knew what they were doing, and the OTA was often the only free source in the state they were not too scared to consult.

CREATING RELATIONSHIPS

Around the time that the Reinvention push brought about EPA's Common Sense Initiative (1996), many states were also grappling with the CSI's aims of governing "cleaner, cheaper, smarter." OTA was developing guidance for industry sectors, in coordination with other relevant agencies, and notably, with the affected community. The first effort had involved printers, and because OTA had hired the former head of their trade association, George Frantz, we played a very large part in this effort. Printers came to the meetings because they knew and trusted George. I have to commend DEP for its wisdom in working this way, in recognizing that nothing good would come of their efforts unless they did this. But I have to comment that this lesson was forgotten partway through, when during one meeting a high-level DEP official announced that DEP would launch the project with eighty inspections, so that it could have a baseline from which it could measure progress in compliance. The head of the printers' trade association blew up and walked out, with every company representative following him. George and I ran after them and caught up with them and got them to come back to the room, giving them our assurances that we would make sure their viewpoints on this matter were heard. (We had gotten them to come to the meeting with DEP.)

As a result, the DEP came to understand that launching a project they claimed would make regulation easier with a blitz of inspections was actually not the way to go. If we hadn't had George on our side, the project would have died right there. The Printers' Project was the model for what became known as the Environmental Results Project (ERP), applied with great success to several industry sectors, and then promoted and taken up by other states as a way to reduce regulatory burden and increase compliance rates, a claim supported by years of data. (The baseline inspection blitz was not necessary.)[8]

---

8. See, for example, the ERP Consortium http://www.erpstates.org/ founded in 2006, or the five-states ERP for autobody shops, 2013: https://www.in.gov/idem/ctap/files/ctap_auto_collision_erp_region_5.pdf.

Not once, during the many presentations I saw on ERP, was there recognition of the role played at the beginning by people already trusted by the regulated community. I don't think people knew or understood. But it was sine qua non—without which, not. This is one of the main points of this book. If not for the relationships of trust that people like George Frantz could establish, it wouldn't have worked.

After the success of the Printers' Project, George moved on to develop relationships with dry cleaners, which then became the basis of one of the most successful ERPs. In 1989, I had launched an effort to reduce the discharge of silver-laden wastes, promoting methods of recovery, which formed positive relationships with photodevelopers, on which another ERP was based. I had also presented at the association of dentists on the necessity of addressing mercury—which was something no one wanted to talk about at first, but we were able to get the discussion going. These and other similar efforts paved the way, enabling DEP to garner the necessary cooperation. EPA promoted ERP because data showed it worked—compliance rates rapidly increased, and people who were not even in the system were now known and could be monitored.[9] But none of it would have been doable if a trust relationship had not been created, and it is one of those stories that ought to be known—how the offer of assistance was necessary to create the relationship with the regulated population, and how getting to know the regulated population enables the appropriate reinvention of the regulatory agency.[10]

HOW IT WORKED

So, what is ERP? ERP is a simplified version of the law that has a self-reporting mechanism. Companies in an ERP sector receive a little booklet that stands in for all the laws they have to obey, and which educates them about safer alternatives they should consider. It is used for small business sectors, not large corporations that have the money to afford lawyers and staff to comb through all of the dense regulatory text that Al Gore needed forklifts to

9. Credit goes to MassDEP's Steve DeGabriele for many years of concentrated effort to prove this, to Terry Goldberg of the Northeast Waste Management Officials Association for bringing it to regional attention, to Beth Termini and others of EPA Region One's Assistance and Pollution Prevention Division for bringing it to national attention, and for Roy Crystal who was also key to the success of the precedent Printers' Project and the later photo-developers' projects.

10. The Massachusetts Water Resources Authority, the organization of Greater Boston communities, was also at this time creating new tools of governance to work better with larger populations of small dischargers, such as the general permit. Its Toxics Reduction and Control program was a frequent partner with OTA and DEP.

pick up. The environmental protection system was at first developed for that entity and needs to evolve to better suit the smaller.

You go through the booklet and you check off if you are doing what you are supposed to do, and you send it in. Every now and then, the agency checks on a statistically valid sampling to see if people are telling the truth. It's the same model as our tax system. If your auditing is sufficient and there is punishment for lying, people will comply. And the ERP booklets were like filing an EZ tax form. After several years of doing the program, MassDEP was able to show EPA not only that companies were complying, but the agency had a much larger set of companies in its radar than before—they weren't hiding out anymore—and their compliance rates had gone way up, compared to inspections before the project.

The role of OTA was not only critical in getting businesses to the table, but to get DEP to listen to them during the program design process. MassDEP, like any government agency, goes through some changes when administrations change. During one period, a somewhat arrogant official was in charge of ERP for a while. I reached out to him, thinking that it's always best to try to create a relationship, even though I had been warned about him, and asked him to come present the plans for ERP at a meeting of an organization we had formed, the Northeast Business Environmental Network (NBEN).[11]

NBEN was a unique entity. It was a group for any kind of business, to discuss their environmental issues. NBEN (first known as the Merrimack Business Environmental Council, because it was formed with New Hampshire on the banks of the river that divides the states), was one of those magical ideas you try out with no evidence that it can work but it miraculously does. We simply issued invitations and companies came. Back then, we had a thriving electronic industry in the area, and these companies, once thought of as "clean," were actually using toxic solvents and facing hazardous waste and air regulations, but they were also stocked with very smart people. Hewlett-Packard, for example, responded immediately and helped bring in several others. At my first meeting we had more than thirty companies, ready to share information with each other, and apparently not worried about OTA being there. We "spun off" NBEN as its own organization, and it was able to hire and keep staff for almost two decades before closing shop. (A similar effort in Central Massachusetts (CMBEN), never incorporated or charged fees, and

11. "We" were George Frantz, Cynthia Barakatt, and I, of OTA, supported by Barbara Kelley.

it's still going strong).[12] One of the things I did was to bring in government officials to explain the rules. It was interesting how hard I had to work to convince some of them to come. They were afraid of talking to companies. Their typical experience was to hold a public hearing on proposed regulations, as required under the Administrative Procedure Act (or the state analogue), and then to bear up as members of the regulated community unleashed angry diatribes at them. I had to assure some enforcement officials that this would not happen, and if it did, that I would intervene to help. Generally, it went very well, with companies happy to be able to ask questions and present their views, and the enforcement people went away pleased and willing to come again. Ice was broken.

THE VALUE OF LISTENING

Only one meeting went badly, and it was the one with that arrogant guy. I will never forget the look he provided the gathered business leaders. It was an attitude of calm acceptance of whatever anyone would say, which felt like indifference. He was sure of what he was saying and he was going to say it, and he was sure it was good for them, and he just did not hear anything they said. As soon as I saw this happening, I knew I had made a mistake reaching out. I have only very rarely felt this in my career.

He explained how ERP would work, and when he told people that they would be self-certifying that they were in compliance with the rules described in the guides, hands shot up. He was asked, "But what if I'm not complying?" He then proceeded to repeat himself, without answering the question. Hands shot up again, and the same question was asked. He then proceeded to repeat himself, and it was obvious now that he was not even trying to listen to what people were saying. Hands shot up again, and the room started to rumble with

12. CMBEN's meetings have been organized by OTA's Marina Gayl, but she has retired as this book is being written. It is that simple act of organizing the agenda and getting speakers that kept the group going for decades. When a government official does that, and the local chamber of commerce gives you a place to meet for free, there is no fee necessary, officers, no incorporation, no staff, and it is easy to keep going. I formed a third BEN in South Central Massachusetts with the help of Hyde Tools' Doug Devries, but he banned me after the first meeting because other members were still deep in suspicion of any government official. Doug died not long after, a great loss to pollution prevention, and the South Central BEN died with him, as far as I know. It is possible it survives and just keeps out of sight. A fourth BEN in North Central Mass operated for a while after the local chamber of commerce agreed to let me invite people there for that purpose, and although the businesses welcomed the events that explained existing and coming rules, it needed someone like me to continue it after I left government service, and it ended. An opportunity exists for BENs anywhere, and because the concept was produced for public use, it is available to anyone to replicate. In all areas businesses need to come together to hear from government about requirements or local environmental issues, and to get information about actions you can take and to discuss all this in a joint project of understanding and improving our need to appropriately respond to the challenges facing us.

anger. People were now absolutely furious with this man. He didn't seem to notice or care. I put up my hand, and I rather think when he called on me that he expected me to tell everyone to calm down and just accept his non-answers. Instead, I clarified the question so that there was no way he could fail to hear it. I handed it to him on a platter so he could say, "we'll look into it" or "I get you. We'll take that into consideration." But he didn't pick it up and reiterated his answer.

THE SYSTEM MAY NOT REWARD YOU: THE WORK IS WHAT'S IMPORTANT

When I got back to the office, I was told that someone high up in the system was very upset with me. I was told that it had been reported to her by this fellow that I had invited him to a meeting in order for him to be mocked and shouted at, and that when the audience became rude toward him, I did nothing to help. I was not given a chance to talk directly to this person and explain things, but had to tell my side of the story through my direct superior, who had to then go defend me. It was this sort of thing that she said tired her out when she retired—we all thought too early.

I now think of these higher-ups who would not talk directly to me as being afraid of me, but I could not at the time, and wondered if I should be afraid of those people, though they were silly. I was lucky to only have to receive doses of this, the kind of thing that makes people not want to work in government. I am fortunate that I could survive and persist through, in order to have a career of developing innovative programs. This was not a terribly hard moment, but one that illustrates the agency problems that can stem from the top. The higher-up could have simply picked up the phone or emailed me.

At a meeting of interested parties the DEP then convened environmental consultant Andrew Irwin and I, and we developed and proposed a concept called "Return to Compliance," which DEP adopted. The final version of the ERP program included this: if you can't say you are in compliance with a requirement, you put in information about when you will get yourself into compliance. You get a grace period. This was the grace note that the program needed to make its music work. It was critical to the program's acceptance by the regulated community.

LEGAL CONTEXT REMAINS FIXER-UPPER

ERP is a way of simplifying compliance, educating people about cleaner alternatives, and getting everyone who should be in the system out from their hiding places. It has been enormously effective. But all during its development, a question plagued us all: What about all those regulatory provisions that were left out when we simplified the law? It took EPA years to decide that they wouldn't make a fuss about that, and would allow DEP to do its experiment, and then to spread the model to other states. But this is still an unresolved question. The states, delegated by EPA to conduct the federal programs, are with this program selecting only some regulations companies need to follow. It's a good idea, for small businesses, to narrow things down to a doable level. But that selection better be good. And like the zero-d wastewater conundrum, there's a faulty legal context. It might leave the companies open to suit by someone else. What if a subsequent administration decides not to exercise forbearance, and pulls out some rule that wasn't in the guidebook? We'd need to have agreement between states implementing and feds overseeing programs, and like the zero-d wastewater permit, we don't want memory of that agreement disappearing.

What's clearly necessary is for Congress to recognize that small businesses do not have the same resources as large, and that such simplified summaries have been shown to improve compliance and performance, and that the approach is better than what came before. With continuous review and assistance to the states in developing such simplified sets of rules, the necessary issues can be covered—better than before—and the law can be changed so that the approach is definitively authorized and no one is left out there potentially vulnerable to challenge, after having tried to do the right thing.

ANOTHER APPROACH TO RELATIONAL REGULATION

While we're waiting for things to happen, we have to keep trying to move the system. Assuming that the vast body of rules suitable for a forklift or two would remain in place, I decided to use the CRASH Course Autobody project that George Frantz was overseeing to demonstrate a new approach, which I called "Pre-contact Good Faith."

This was during a time when EPA was beginning to develop new approaches to enforcement, recognizing that the old practices were too limited, too inflexible, and not always fair. The agency was developing environmental management system and self-reporting policies, by which companies that adopted a regular practice of self-auditing, upon finding violations, could report them, and gain immunity from enforcement. Think of it as another form of grace, as one is forgiven after confession in the Catholic religion, or one atones during Yom Kippur in the Jewish, in order to "Return to Compliance," get a fresh start, not unlike when a nation that has committed great sins engages in Truth and Reconciliation. EPA's policies did not apply to grave violations, so in Region One you could call Dwight Peavey and he would not ask you who you were, and answer your questions about the forgiveness, to find out for sure that you would get it, before you turned yourself in.

It is unfortunate that some right-wing politicians seized on this to demand or try to institute complete immunity, but this reckless movement did not succeed and the self-reporting *privilege*—not immunity—worked to give companies the confidence to remedy many violations, instead of hiding them and having to live with the fear that they would be found out.[13] Such approaches need expansion, because they work, but utilization is far less than it could be. It worked in Region One, but Dwight has retired, and his extensive knowledge of the law went with him. There are other forms of enforcement forbearance that can encourage companies to come forward, and be restored as members of the community in good standing.

But all the alternative approaches take time to develop and implement, and talented staff—who know enough to see when companies are not acting in good faith. In order to design a program of forgiveness that works, it is necessary to know if a company has just found one or two violations recently, or if they have had hundreds, for years. It is necessary to know if a company has just made a mistake, or just doesn't care. We need to distinguish between those who are trying, or new and ignorant, to those who knowingly have avoided responsibilities for years and have benefited from that and competed with those who comply. It is imperative to treat members of the regulated population as individuals to the best of our ability, and try to determine "pre-contact" good faith.

Pre-Contact Good Faith

My concept was that the CRASH Course guidance (not at that time part of the Environmental Results Program) provided a means of documenting good faith actions *before* a company came into contact with the enforcement agency. Enforcement Response Policies (also using the acronym ERP) are used by EPA to determine penalties. They evolved to give credit for good faith, so that companies who cared would get lesser fines than companies who

13. With a privilege, the company could withhold information about it, but that veil could be pierced if needed, because of the gravity of harm, willfulness, repeat violations.

did not recognize their duties to operate responsibly. But EPA's Enforcement Response Policies only pertained to "post-contact" good faith.

If you are caught by EPA in violation, and you immediately take steps to cure the problem, you get something off your penalty. If you then settle with EPA instead of going to court to contest the penalty, you get something off your penalty. When I later worked at EPA as an enforcement attorney, I discussed this issue with a fellow attorney, one who I think may have never worked anywhere else. He said, "We assume everyone is in violation until we catch them." This summarized what I was going up against when I wrote up my concept in an article for the Environmental Law Institute, titled "Good Faith as a Fundamental Principle of Relational Environmental Governance."[14] I wrote that we needed not just to reward that post-contact good faith, what a company did after being caught by an agency, but pre-contact good faith as well. If a company could show a demonstrable commitment to environmental performance before contact with the enforcement process, it should enjoy a better outcome than a company that had not, even if it was found in violation of something. If we could do that, it would encourage the right behavior by companies not yet visited by the authorities.

It was a commonplace, because there were so many rules and many companies didn't have a handle on all of them, for company staff to tell me, "Everybody's in violation of *something*." And in my work, visiting hundreds of facilities to help them find violations before an inspector did, I found that this was generally correct. It got better over the years, but when I started, violations were widespread, though there were always some very clean facilities, where owners and operators cared.

> Precontact good faith creates a different enforcement response for company A, which is making an attempt to do the right thing, than for company B, which does not care. Postcontact good-faith rewards cooperation after the inspection has occurred and the penalties are being assessed, and unless the same companies who try to do the right thing on their own are the same ones who tend to cooperate, it does nothing to distinguish between company A and company B.[15]

If we could distinguish between companies that are trying, and companies that simply blow things off, and then reduce penalties for the first, we could encourage more companies to be like company A. If we didn't adopt this course of action, then what were we saying? We were saying all companies were violators until we catch them, as my colleague described enforcement practice. This discourages companies from doing the right thing. Why should

14. *Environmental Law Reporter*, February 2, 2002, 32 ELR 10219, http://www.bu.edu/earth/files/2012/08/goodfaith.pdf.

15. Ibid., 10225.

The Commonwealth of Massachusetts
CRASH (Collision Repair Auto Shop Help) Course
for Compliance and Pollution Prevention
Manual

October 15, 1998

Dear Auto Body Shop Owner or Operator:

We recognize that for a small business person, it can be difficult to understand and follow the many local, state and federal rules designed to ensure safety and a clean environment. That is why we have produced this manual: to help you understand your obligations; to help keep your shop safer for your employees, your neighbors and the environment; and to help you stay in compliance with the law.

The enclosed manual provides information you need to understand and to meet your environmental obligations. The greatest benefit to you and the environment will come from preventing pollution. In this manual you will find not only the rules which apply to your business, but tips on best management practices which will help you go beyond compliance and may save you money as well.

By using the self-assessment checklist you will be able to determine how well you are doing. Used properly, it can demonstrate to an inspector your good faith effort to comply. The "Tool Box" will give you additional resources, including state and federal small business compliance incentive policies which may help you avoid costly penalties.

We encourage you to use this manual with your employees.

Trudy Coxe
Secretary, Executive Office of Environmental Affairs

John P. DeVillars
Regional Administrator, Environmental Protection Agency

David Struhs
Commissioner, Department of Environmental Protection

Scott Harshbarger
Attorney General

Robert J. Prezioso
Deputy Director, Division of Occupational Safety

The Commonwealth of Massachusetts

Figure 9.1. CRASH Course for Compliance and Pollution Prevention Manual

they spend so much time trying if when they have one little slip up, they will be treated just like the jerk down the road who pollutes with abandon? This failure of what I called "Relational Environmental Governance" was a huge missed opportunity.

The CRASH Course gave auto body shops a guide with checklists. For a couple of years, I pitched to enforcement officials this idea: that if an inspector came and saw that the shop was using it, and the checklists had been filled out, and what he or she saw in the shop corresponded with the paperwork, that could be evidence of good faith. They worried they would be giving something away prematurely. I replied that this needed only be a promise on their part to *consider* the evidence—so it need not create any barriers to enforcement when it was needed. This finally worked and I gained the support of the top officials of EPA, DEP, and other relevant agencies. In my office, I have on the wall this piece of paper (shown in figure 9.1), with their signatures.

The work of more than two years was in that one sentence:

> *Used properly, it can demonstrate to an inspector your good faith effort to comply.*

This policy, when announced to auto body shops, was very effective. Shops filled out the guidance, and put the checklists up on the wall, and made sure that what someone saw in the shop corresponded to what was on the wall.

FADING OUT

But this story has a mixed ending. The practices of the CRASH Course were more widely adopted, and the ways of cleaner operations advanced. But after about ten years, it had been completely forgotten. All the administrators who signed the paper above were gone and when I would ask DEP inspectors if they were asking shops for the CRASH Course documentation, they would say, "huh?" The project did work at one time and for a while. But these things have to be kept up by successions in leadership. When you look at general literature about improving governance, it is not hard to find the comment that there needs to be more continuity between administrations. The arrogance of some new high officials, who think they don't even need to learn about what has come before, that it is now *their* time, and *they* are the ones with fresh ideas, the old guard has clearly failed, is unfortunately too common. (I've seen it in both parties.)

How lovely would it be if we had the practice by which new members of an administration learned about what has come before, and wanted first and foremost to preserve whatever advances had been made, to continue that which is working. They would more likely want this if they felt we who are watching them would want this.

PRESERVING THE FOUNDATION

This approach of recognizing and rewarding pre-contact good faith differs from ERP in that it does not eliminate all the other rules that pertain. It did provide a simplified guide to compliance, but while ERP actually said this is all you have to do, the CRASH Course did not say that. Instead of saying the equivalent of "we have just eliminated a whole raft of rules passed by Congress that we as a state promised to enforce when we took delegation of the law,"[16] it essentially says "we will use enforcement discretion to focus on these particular rules." This seems more accurate and a more legally clear means of simplifying the rules.

Both approaches reflect a more relational kind of governance, in which agencies recognize the realities of the regulated community. Both reflect knowledge and caring about the situation small companies are in, and both work to improve the ability of those companies to exercise responsibility toward the rest of us—as if we are all in relationship with each other and the world. We need to pull our heads up out of the fields in which we have specialized, and see each other as we are. One of the reasons I wrote that article was to tell people about the project. But another was simply to see in print the term "Relational Environmental Governance."

Cynthia Giles of Harvard Environmental Law, who has had extensive experience at EPA, writes in *Next Generation Compliance: Environmental Regulation for the Modern Era*[17] that noncompliance is really quite high where we know what it is, but in many cases we don't even know what the rate is. She calls for better regulatory design instead of just applying force, as the stories I tell above try to show. As a prime example she compares the Clean Air Act's Acid Rain regulations with the New Source Review regulations. Under the Acid Rain regulations we had continuous monitoring, which supplied good data and the determination of compliance was simple, and that made compliance the cheaper option. But under the New Source Review the agency had to do a lot of work to determine compliance, and the compliance was expensive. The cheaper option then, for the regulated community, was to see if you could get away with not getting caught. When Giles was head of enforcement under

16. The major environmental laws work by delegation—EPA sets standards that the states must follow and when they have shown they are in compliance they receive money from the federal government. Thus the federal environmental laws are primarily, but not entirely, implemented by the states with oversight from EPA.

17. http://eelp.law.harvard.edu/wp-content/uploads/Cynthia-Giles-Intro-FINAL.pdf.

Obama, she promoted the use of advanced technology for better compliance monitoring—for example installing "solar powered continuous monitoring devices that upload via cell phone technology to agency computers."[18]

IMPROVING COMPLIANCE MECHANISMS

In 2017 Ken Zarker of the National Pollution Prevention Roundtable informed the board that we were invited to send a member to participate in a "regulatory negotiation" concerning how recycled inorganic byproducts are regulated under the Toxic Substances Control Act (TSCA). I volunteered. I might have been the only person there not being paid in some way for my time, but it was a great opportunity to work with a law that I think of as having great unused potential, that I had made a special focus of during my career, marshalling comments on state access to TSCA data and against preemption of state toxics laws. Companies claimed the reporting required by TSCA's Chemical Data Reporting rule was burdensome, but we environmentalists pointed out that some of the substances under discussion were extremely dangerous, so why should EPA give up information about them?

Take a look at the reporting form—the Form U (shown in figure 9.2).

Manufacturers over certain thresholds have to report this. They have this information, and knowing that they have to report it helps keep safe operations in place.

Then on the next page of the Form U you have what is shown in figure 9.3.[19]

This pertains to all the downstream processors of the materials the manufacturers sell to them. The manufacturers are supposed to put in this information. It would be good information for society to have—whether the chemicals are in products intended for children, for example. But do manufacturers have all that information about the customers to whom they sell their product? We asked the EPA attorneys about whether manufacturers, who were asking us to reduce the burden of reporting, had to get on the phone to their customers. They said the company just had to search its own data, for that which was readily obtainable. Chemical data reporting (CDR) guidance states:

> there is not a need to conduct new customer surveys for purposes of the CDR. As described above, however, existing customer survey data may nevertheless be "known to" the organization.[20]

---

18. https://www.epa.gov/sites/production/files/2014-09/documents/giles-next-gen-article-forum-eli-sept-oct-2013.pdf.
19. https://www.epa.gov/sites/production/files/2020-03/documents/cdr_primary_manufacturer_submission.pdf.
20. 24.1 at https://www.epa.gov/chemical-data-reporting/completing-form-u.

Section C. Manufacturing Information

Non-Principal Reporting Year(s) Production Volume Information **CBI**

Calendar Year 2018

Calendar Year 2017

Calendar Year 2016

Section C.1 Manufacturing Company

Report CY 2019 Production Volume and Related Information	**CBI**		**CBI**
Activity	**Manufacture:**	Activity	**Import:**
Domestically Manufactured		Imported	
Imported Chemical Never Physically at Site		Volume Used on Site	
Volume Exported		% of Total Production Volume (by weight) that is manufactured as a Byproduct	

Report Exposure Related Information	**CBI**		**CBI**
Number of Workers		Max Concentration	
Is the chemical being recycled?			

Report Physical Form for 2019 Production Volume

Form	CBI	% Production Volume	CBI
Dry Powder Form			
Pellets or Large Crystals			
Water or Solvent Wet Solid			
Other Solid			
Gas or Vapor			

Figure 9.2. The Form U

Form	CBI	% Production Volume	CBI
Liquid			
Not Known or Reasonably Ascertainable (NKRA)			

Section D. Processing and Use Information

Section D.1 Industrial Processing and Use

Process and Use Information	**CBI**		**CBI**
Type of Process or Use		Sector(s)	
Function Category		*Function Category (Other)*	
Percent Production Volume		Number of Sites	
Number of Workers			

Section D.2 Consumer and Commercial Use

Product Category Information	**CBI**		**CBI**
Product Category		Function Category	
Consumer or Commercial		Used in Products Intended for Children	
% Production Volume		Maximum Concentration	
Number of Commercial Workers Reasonably Likely to be Exposed			

Part III. CONFIDENTIAL BUSINESS INFORMATION SUBSTANTIATION

A person may assert a claim of confidentiality for the specific chemical identity of a chemical substance as described in § 711.15(b)(3) of this part only if the identity of that chemical substance is treated as confidential in the Master Inventory File as of the time the report is submitted for that chemical substance. Generic chemical identities and accession numbers may not be claimed as confidential. To assert a claim of confidentiality for the identity of a reportable chemical substance, you must submit with the report detailed written answers to the questions from subsection (b) and to the following questions.

Figure 9.3. The Form U

The attorneys said you could say, "Don't Know," if you didn't happen to have the information.

The way we did the negotiation was to talk as a group and then industry would go to one room and the enviros to another, and decide what questions we have and how we feel about what we heard. I shared with the environmental caucus that I was going to propose that we ought to shift the burden to where it needed to be, and have the processors reporting, and that this would reduce burden on the manufacturers. They wouldn't have to even think about picking up the phone. The whole thrust of the negotiation was that the chemical reporting was burdensome.

Amy Kyle of the University of California, Berkeley, said that she thought we could reduce the burden of reporting by using all the modern electronic internet technologies that are now widely available. The form could be much easier to fill out and send in than it was. The two ideas together come to a system in which the form follows the chemical.

You are a manufacturer and you are about to sell chemical X to a customer. You pull up the new Form U and put in the name of the chemical, where it's going, and how much. That's all you would have to do. The first page shown above: you send this into the central chemical tracking database and you send a link to the customer. Then when the customer gets the chemical X shipment, he or she taps on the link and adds the information that *they* have—an estimate of how many workers might handle it, and whether or not it's going into products meant for children.

This system would have much less burden on the manufacturer. It would provide a complete picture of where chemicals are going and into what. We would know where workers are at risk and we would know a lot more than we do now about how to protect children.

I wrote this up as a proposal and delivered it to the entire group when we considered potential approaches to fixing Form U.[21] One of the industry representatives was a young man and he started nodding his head vigorously as I talked, clearly loving the idea of taking away all that annoying and, actually, rather impossible and ridiculous reporting burden from his clients. But then a much older man, who was there as an observer for the oil or chemical industry, got up from the side and walked over to him and whispered in his ear. This was not something that usually happened. Observers generally kept to the side and didn't interfere. Suddenly the young man's hand shot up and he declared before being called on that he opposed my proposal. This was very much out of line and was the only time during the entire negotiation that

---

21. Proposal D at this site: https://www.epa.gov/sites/production/files/2017-08/documents/3-draft_approaches_cdr_inorganic_byproducts_081417.pdf.

anything like that happened. *We were supposed to just listen to all the proposals and then discuss them in our caucuses before commenting.*

The negotiation did not succeed in producing anything. The industry proposals would all have reduced information that the public should have about the movements of dangerous substances. The format of the proceedings did not allow very much personal interaction between the industry and environmental representatives, but I did have some brief discussions with companies that were there who were not from the oil or chemical business, generally much smaller companies. They said, it seemed somewhat sheepishly, that the idea of reducing regulatory burden that I suggested was a good one, but that the group didn't feel that they could agree to it, because processors were not there.

I don't know if this means the claims about reporting burden were a sham, and the real intent all along was to reduce the information the public gets. But I do know that when there are genuine concerns about an industry's products or activities greater awareness of safety is the only sensible route to take, and that begins with knowing the hazards.

It wasn't wasted time for me personally. I felt rewarded by the chance to work on the regulatory puzzle and come up with a solution that felt to me so simple and elegant, showing how we can get better information and lighten, not increase, regulatory burden. For my particular career it was a high point and I get to share it with the world, now, in this book.

There are many things like that—we reduce regulatory burden by making our rules smarter, not weaker.[22] We see the same thing with Cynthia Giles's Next Generation. Put a monitoring device next to the tank and have its data go to the agency, and the minute it leaks the agency knows. The company saves money because it doesn't have to put employees on time-consuming leak detection duty, the problem gets addressed right away, and because it was, there's no penalty. Leaks happen. It's better for all concerned if they get addressed right away. Compare that system to one where the leak isn't detected for months, and the agency can impose a "per-day" penalty. On top of the fine being so high the company also has lost valuable material and perhaps the trust of the community as well.

In 2004, I spent a year at the Massachusetts Department of Environmental Protection working on enforcement policy. I reviewed all the settlements that gave companies a break in penalties in exchange for putting an "Environmental Management System" in place. I found that there was tremendous variety and got permission to write up a guide to how it ought to be done.[23] In that guide I suggested that if the company was willing to install real-time

22. https://www.theregreview.org/2017/11/29/reibstein-improve-regulation-not-repeal/.
23. I must express my gratitude to Ann Lowry for supporting this effort.

monitoring that it should get a big reduction in penalty. It would save them time, save the agency time, and provide assurances that the process that had been emitting, or would emit, would remain under control, and if not, that we would know about it right away. That would be an Environmental Management System worth giving a huge penalty break for.

Companies can recognize that they have a responsibility to keep potentially polluting operations under control and that monitoring is the best way. They can get out of the habitual thinking that it's best to not let the agencies know anything. That's the old way, and if more citizens are aware of the fact that we have a choice of a better way, maybe companies can be helped to come into the modern age. New technologies—nano, for one—have given us cheaper monitoring. With rapid response we can tighten up operations tremendously.

REVIVE THE HUMANE PURPOSES OF ENFORCEMENT

These methods of improving enforcement are rather technical—redesigning the regulations, using monitoring. But there is a deeper change that agencies should implement, and informed citizens should support. It is to utilize the often-forgotten purposes of enforcement. Enforcement is typically thought of as punishment. But it should also be about restitution and rehabilitation. These aims could be primary. The idea of punishment has a primitive aspect—retribution—hit me, I hit you back. It also has a more sophisticated aspect—it makes you aware of what you need to do, and you are rehabilitated.

We should be conscious of the fact that we are playing a game of paternalism here, that we are thinking of the agency as the parent who molds the behavior of a child. The fact is, these are adults we are talking about, and we cannot easily reach their inner selves and mold them. Punishment may just lead to resentment.

In my experience in enforcement, I saw two kinds of defendant: repentant and resentful. About half or a little more seemed to truly regret what they had done and wanted to make amends, and nearly half seemed to just regret having been caught, and some were angry that we had caught them instead of the "guy down the street who's much worse than me!" With far more attention to the question of good faith, pre-contact and post-contact, we can do a great deal to sort these out. Those willing to accept responsibility will settle for monitoring or reporting that provides assurances they are remaining in compliance. They would be clearly exhibiting good faith and deserve a break in their penalty. Similarly, those willing to do something to make amends, what's called a "Supplemental Environmental Project" (SEP), deserve a break. These are established ways of allowing for restitution and rehabilitation. But they need shoring up.[24]

HOW WE DEFINE SUCCESS

When we count up the amounts obtained in penalties and hail that as an accomplishment, we should consider it a failure of our system, that so much punishment was needed. We should worry about how the money gained through penalties is spent. When it is spent on restoring a resource, or teaching others something they ought to know, demonstrating how to make something without using toxic chemicals, we are benefiting the future as well as restoring the violator to the good graces of society. An SEP can do that, while just sending the fine to the Treasury might not.

When I was at EPA, I had to call up a company that was in trouble for excessive air emissions of a toxic solvent. The company president was shocked—crestfallen—to receive the call. He begged me for some way to soften the blow. It wasn't the largest penalty I had the duty to levy, in fact it was relatively small, but he was clearly humiliated and feared the loss of a reputation he treasured as a proud member of his community. I told him, as I told everyone, that he could get the penalty down by doing an SEP.[25] I told him that I would ask the press office not to issue a release on the complaint, giving him a chance to think about what SEP he could do, and then when we did the release, it wouldn't say just that he had a violation, but that he was doing something about it.[26] He was happy to hear about all this but said, I don't have any idea what I can do, for this SEP!

I told how I had come from OTA and had helped companies find toxics use reduction opportunities. I said that he shouldn't interpret what I would suggest as a command, but given that, I could give him some ideas. I made an appointment to come down and see the factory. When I got there, I immediately saw that he had an old solvent degreaser—a huge one—and I told him there was now a whole range of opportunities for degreasing more safely, that companies had been getting rid of those old things for years. They were big emitters and bad for employees too, and there are so many choices for cleaning he could consider instead—he could use an aqueous cleaner with ultrasound, and he could also investigate why his parts got oil on them and maybe

---

24. Attorney General Sessions worked to stop SEPs.

25. Some attorneys did not emphasize the possibility of an SEP, having little experience in managing projects, and wanting to move on to the next case. But I had worked at OTA for several years before coming to EPA. I was determined to do a Pollution Prevention SEP with every case I was handed.

26. This, also, was not something other attorneys necessarily did.

he could avoid the need to clean with any kind of solvent. We worked out a deal in which he removed the machine, stopped using the toxic chemicals, and then no longer had to report under the Clean Air Act, and had a much safer workplace. His penalty was reduced, and the press release included the good parts of the story, noting that it all began with an enforcement action (this is required).

This company president then wrote a letter to the regional administrator (RA) thanking me for enforcing against his company. One of the enforcement chiefs came down to my cubicle to share the letter with me and say that the RA was very impressed and grateful for what I had done. He stared at me. I asked him why he was looking at me like that. He said nothing like that had ever happened before. No one had ever thanked the agency for enforcing against them. I went home pretty happy. But after a while I got the feeling that not everybody felt the same way as the RA. I asked the chief about it. He laughed and told me not to worry. "You did the right thing. It's just that some people wonder if you maybe got too friendly with the guy." Then he said, "It's just not done, that we go to them. We always have them come to us!" The fact that I had visited his factory was a problem. He was supposed to come to EPA offices.

I think that there is no reason to give undue respect to those who violate the rules, and I had one case with someone who badly needed punishment and to be put out of business, for he was harming people. But most of the people I dealt with were not like that. Many of them just failed to appreciate the need for compliance, and there are a lot of reasons that can happen. I'm not saying we need to go easy on people. I'm saying we need to find the right relationship with them and act to prevent violations, just like we can prevent fires, explosions, disease, and pollution. Sometimes that can involve going to see them where they are.

A CREATIVE BUREAUCRACY

There are so many things that can be improved. Some are complicated, and some are easy. One that ought to be easy is disposing of the pernicious form of the "unitary executive" idea. The legitimate form of the idea is that one person is the executive, who gets to declare policy for all under him or her. But that is to be done according to the mission, which includes the opinions of all those in the agency who have been working to accomplish its mission. The pernicious and absurd form simply says everyone has to obey the executive, without the proviso that they must obey when the executive is fulfilling their mission, and ought not to when they are not. In contrast, we need a stronger ethic of loyalty to the people as a whole, not the current person in power or their party. Environmental citizens ought to be able to get on the phone and talk with their public servants in public agencies and not have that line of communication controlled by political appointees, as is now commonplace.[27]

We ought to have government that is dynamic and full of talented people who are free to do their job. We have been gravitating toward something much less capable and much more brittle.

I have seen many creative ideas from the middle, the bottom, and the top. Good ideas come from all levels. Some leaders take credit for what those under them produce. Picture this: a leader says we need to address a particular problem, and staff then meets to generate ideas. The best ideas are then reported up the chain. Unless the key person at each step in that chain has the knowledge of who first came up with the good idea, and the decency to point that out, it tends to get lost, if it was ever noticed in the first place. Too many times in meetings someone has suggested something, and no one says anything, and then when someone else suggests the same or a similar idea, it is suddenly heard. This is an artifact of poor listening, or of not wanting to recognize someone, or of wanting to please others, or of having to hear something more than once before it clicks into place. And it very often occurred that the idea was not heard until someone important expressed it. Then the loyal followers thought it was terrific. This is dismaying to see, but it is even more dismaying to realize that when this sort of thing happens a lot people who don't get heard stop offering their good ideas.

THE UNITARY EXECUTIVE "CAPTURES" AGENCIES

When George W. Bush was elected the second time, he announced that he would be reaching out to those who agreed with him. This was such a sad moment. He needed, as leader of the entire country, to reach out to those who disagree with him. But he was so immersed in partisanship that he didn't understand this was his duty. His vice president, Dick Cheney, stated repeatedly that he believed in the "unitary executive," the idea that government must speak with one voice. Simple top-down command obedience is essentially the fascist and not the democratic system. It destroys the creativity that government must develop.

The job of the executive is to faithfully execute the laws that have been passed by the legislature and signed into law by previous executives, and

27. Another manifestation of this inappropriate executive control (engaged in by both parties) is when an agency website does not provide the contact information for mid-level staff but only the top officials and communications specialists.

those laws reflect many purposes and require many voices. Executives do their work when they ensure people under them do what they are supposed to do to serve the people, not to serve the executive's party's platform. The concept that everyone follows the latest leader, eliminating diversity of views, preventing staff from correcting the mistakes of leadership, heightens the danger of such executive power being misused, and curtails the ability of the agencies to respond directly to the needs of the people. It displaces trust from the deliberative process to the leader of the moment, which goes against the founding spirit of the country that replaced a king with the People.

When administrations change, each succeeding executive excessively believing in themselves rather than the combined talents of all the staff of all the agencies may feel that they properly and righteously institute changes in accord with their sense of the mission. A lack of curiosity and respect for what already exists in the agencies can introduce a ferocious instability in the system, as we saw when the disrespectful Donald Trump reversed everything that he could of what Obama had put in place. Now, President Biden has to reverse those reversals. Had the concept of the executive's will not become so determinative, we might have saved even more of what the far right has disparaged as the Deep State. Thank God for its depth, or even more would have been lost in the storm.

REORIENT THE AGENCIES TOWARD THEIR ORIGINAL MISSION

When I entered government service, it was during the administration of Massachusetts governor Michael Dukakis. He was known as a "goo-goo"—a good government man, a leader who respected those who served under him, and wanted them to be creative. He would visit agencies and walk around and visit cubicles. I used to receive calls from legislative staff, asking me to review bills. I used to receive calls from journalists, and I would write articles for publication, and nobody stopped me, nobody said you can't do those things. But as time went on, and successive Republican administrations came and went, those freedoms were lost. The message was controlled. I was not allowed to respond to calls from the legislature or journalists until I got approval. I regard this kind of control as a violation of the separation of powers, a usurpation by the executive. Agencies are created by the legislature. Our bosses are the legislature and the people they serve. The taxpayer was paying my salary. They and the representatives and journalists seeking information had a right to speak directly to me.

When, finally, after many years a Democratic administration came in, that of Deval Patrick, I breathed a sigh of relief. But while it was easier to get approval to respond to queries, the practice of having to seek approval was not changed. I chalk this up to the fact that everyone who came in was so young that they didn't even know how it used to be. But it also served their purposes, as our politics has become so divisive that controlling the message now seems necessary, in order to avoid letting the other side have any excuse to blow things up.

I think, at this point, of a journalist named Howie Carr, who regularly wrote about "hacks and flacks" in the *Boston Herald*, promoting the idea that he stood up for the regular people in exposing the faults of government. But on one occasion, he wrote about someone I knew, someone who I knew was making important contributions, and I saw not only that Carr's simplistic cynicism was harming him, but all of us. When you run down people in government, you make it a place where the talented don't want to be. You make it easier for the public to approve of a "strong leader" who will supposedly force the agency to do better. You make the agencies close up even more.

It is the job of the executive to ensure—and help—staff do their jobs. When people in the legislature can talk freely to experts in agencies, bills will be better. When the public or the press can get immediate access, they will learn what actually does work, and can develop an appropriate and measured faith in the parts of government that are responsive to them. Opening up government will require cooperation from the top, not ceding control to it.

A MID-LEVEL STAFFER AWAKENS EVERYONE TO ELECTRONIC WASTE

I witnessed a powerful example of the value of empowering staff during the Common Sense Initiative. Region One's administrator, John DeVillars, provided strong support to the work of a staff member named Mark Mahoney, who had been facilitating discussions on pollution prevention for years. Mark began one meeting with a proposal for something to talk about: the problem of electronic waste. It was 1997. Mark asked the room, filled with people in or concerned about the electronics industry, "What happens to all the computers people are starting to buy, all the electronic games, the cell phones?" Nobody seemed to have given this question any thought, not the industry or environmental advocates or officials.

It wasn't long before we realized that Mark had put his finger on something enormously important. We found almost no research had been done, but that which we could find revealed huge quantities of problematic waste.[28] We started analyzing what was in it, and realized that there were some very problematic constituents. Earlier in this book, I discussed the lead in glass issue, but in computers there was also mercury, chromium, polyvinyl chloride, and beryllium,[29] and because there was also gold and copper, people had started to break the old equipment up to recover these valuable materials, sometimes burning everything else off and causing serious toxic contamination. We funded some collections to find out what people would bring and we prompted further study and then we reported our findings. I will never forget how some people got angry. They accused us of wasting time on a trivial issue. One of those who did became convinced shortly afterward that he was wrong, and became a major proponent of acting on this problem. We formed an Electronic Product Recovery and Recycling Roundtable (EPR2), the first national organization to address the issue, and held two national conferences. At the first, political commentator Norm Ornstein said that we were one of those "points of light" that George H. W. Bush talked about. Both Mark and I were recognized for the work we had done to create the EPR2 Roundtable.

CONFUSION BECAUSE POLITICS CAME FIRST

But then, in my state, policy warfare broke out. The head of the DEP, a Republican, decided to deregulate CRTs—computer and television screens—as a hazardous waste. At the same time, he banned them from disposal. This was a good result, but an unlawful method. I got a call from a top regional EPA lawyer—during a Democratic administration—asking me, "What is he doing? He doesn't have the power to deregulate a hazardous waste"![30] I was asked to communicate this objection to the Republican Commissioner, which I did. We actually had a great relationship. I enjoyed talking to him. I said it was great he was forcing recycling, but that the EPA said he didn't have the authority to do it this way, and wanted to work something out.

Unfortunately, he didn't seem to want to work anything out, but seemed to want to have a fight. He went down to the second EPR2 Roundtable and denounced EPA. The next speaker said, "I thought we were here to talk about ecosystems, not egosystems." I, unfortunately, was not there, because I had been unfairly blamed for causing the ruckus. Before I could explain that

28. One prominent example was a collection of electronic waste proposed and overseen by then Somerville Massachusetts's environmental coordinator, Hilary Eustace.

29. Berylliosis, a dangerous lung disorder, was of particular concern.

30. Jeff Fowley now teaches environmental law at Boston College.

I was merely transmitting what EPA had told me, I was instructed to stop working on electronic waste. I was generally extremely lucky in my career in bureaucracy, and was empowered to design project after project. I was mostly blessed with the support of my superiors and the freedom to create. But not by all. A higher-up wanted to stop the fuss and without caring about the facts, thought the best way was to simply push me out of things. She rushed to judgment, never picked up the phone to speak with me.

ABANDONMENT

George W. Bush's EPA then abandoned the EPR2 Roundtable. EPA had worked out a cooperative agreement with the National Safety Council to operate it, but that wasn't really a good fit, and after only two years they had dropped it, and EPA did nothing to keep it going. We had built a new national movement to prevent or at least manage the immense mountain of electronic wastes, and it was abandoned. In developing countries there are places where children have high levels of mercury and other toxic poisoning because they live in dumps where people burn electronics to get the bits of metal that will help them get by. We love our electronics but because of the lack of a national movement to control their fate we have let this happen. Yes, the movement Mark began did lead to Best Buy and others taking back electronics, and product stewardship laws in some states, and the establishment of many efforts that have made a difference. But it could have been far more effective with continued national leadership.

Had we kept up the momentum, we might have had in place by now a national requirement for take-back. But most ended up only taking back their own equipment, and only when it made business sense to do so by their lights—this was called Asset Recovery—as many components could be refurbished or remanufactured, saving money on raw materials and making things from scratch. That some companies take back some electronic waste is not enough. EPA's website refers interested parties to West Virginia's Electronics Recycling Coordination Clearinghouse, which has maps of state legislation, showing that there are twenty-five states with laws about electronic equipment, only some of which ban it from disposal in the ordinary trash.[31] Meanwhile, the Silicon Valley Toxics Coalition tells us that

> The US Environmental Protection Agency (EPA) estimates that the stream of e-waste is growing two to three times the rate of any other source of waste. Only 15–20 percent of e-waste is recycled, and, according to the EPA, the "vast

31. https://www.ecycleclearinghouse.org/maps.aspx.

majority" of that waste is exported. California alone exported an estimated 20 million pounds of e-waste in 2006.[32]

THE REAL UNFUNDED MANDATE

Those exports may be to places where the waste is smashed, freeing up toxic particles to be breathed, or burned to recover the metals, creating toxic smokes.[33] A few dozen nonprofit organizations have banded together to form the Electronics TakeBack Coalition, promoting Extended Producer Responsibility and the green design that goes with it.[34] If you are a maker of electronics and you know you will have to take it back at the end of its useful life, you will design it so that you can easily dismantle it and reuse what's still valuable. If you don't have to take it back, you can skip that work. Scott Cassel, founder of the Product Stewardship Institute, which promotes takeback of paint, mattresses, and many other items, calls the current system that imposes on municipalities the job of managing our wastes, an "unfunded mandate." In contrast to the use of that term by conservatives to refer to requirements to test drinking water, this is a mandate imposed not by government, but by the fact that the wastes must be managed and the businesses that make money from producing them are not contributing to the cost of management.

The vision of the Electronic Product Recovery and Recycling Roundtable was not explicitly focused on regulation such as a national take-back requirement. But it was a foundation for discussing what was needed, and that is the eventual place we must land. I shared a proposal with business representatives at the Common Sense Initiative, a little schema I worked up, circular in nature, with a small fee at purchase covering recovery costs, and how it could be structured with choice, where the producers take responsibility for their stuff or contribute to the government effort that fills the gaps. They found it reasonable, but one gave me to understand that the prevailing thinking in the industry was how to prevent a law in the US like the European post-use requirements (specifically, what is now the Waste Electrical and Electronic Equipment Directive).[35] I was stunned by his honesty and grateful. He didn't

32. http://svtc.org/our-work/e-waste/.

33. https://www.smithsonianmag.com/science-nature/burning-truth-behind-e-waste-dump-africa-180957597/#:~:text=During%20the%20last%20decade,%20some%20of%20the%20world%E2%80%99s,narrative%20that%20resonates%20strongly%20in%20a%20technology-obsessed%20world.

34. http://www.electronicstakeback.com/.

35. the European Commission's Waste Electronics and Electrical Equipment (WEEE) Directive, entered into force in 2012, requires those who make the products that become this waste stream contribute to efforts to recover it, https://ec.europa.eu/environment/topics/waste-and-recycling/waste-electrical-and-electronic-equipment-weee_en.

say it, but I assumed that people in his industry might be thinking that by working with us on fostering *some* recycling, they might avoid regulations that required it. I often wonder if any of the people from the electronics industry who worked with us have, since that time, worked for effective, universal rules, or in some way to counter that sorry ethics, that passive aggressive behavior of the industry. From the perspective of Relational Citizenship, corporations should show respect for the government's job of addressing real, common needs, especially when they are imposing harm and costs on the rest of the world. They have a responsibility to submit to this reality in some way, and should seek regulations that make sense, not their absence. Consumers can demand this behavior but first must be aware of it.

WE CAN BEGIN AGAIN

It was a beginning that can be begun again. The issue has been kept alive—Chris Beling, for example, in Region One EPA continued to work on it, as did several states. A CT memo from 2007 on setting standards for e-waste recyclers noted that "There was a consensus that the use of prison labor should not be allowed."[36] Patty Dillon, who was part of CSI, went on to help drive the development of EPEAT, the Electronic Product Environmental Assessment Tool,[37] which became a global standard for ecolabeling IT, managed by the Global Electronics Council.[38] There is much happening in this space. But government could have been playing a much stronger coordinating role, moving the industry forward much more rapidly than it has been able to evolve thus far.

Describing his presentation "Ewaste—the Underestimated Challenge," for the 2020 (the third) annual expo of the ewaste industry, Dr. Ruediger Kuehr[39] wrote:

> A record 53.6 million metric tons (Mt) of electronic waste was generated worldwide in 2019, up 21% in just five years, according to the UN's Global E-waste Monitor 2020, released recently. The new report also predicts global e-waste—discarded products with a battery or plug—will reach 74 Mt by 2030, almost a doubling of e-waste in just 16 years. This makes e-waste the world's

---

36. https://portal.ct.gov/-/media/DEEP/reduce_reuse_recycle/electronics/minutes111407pdf.pdf.
37. https://www.epa.gov/greenerproducts/electronic-product-environmental-assessment-tool-epeat.
38. https://cacm.acm.org/magazines/2020/7/245692-reducing-and-eliminating-e-waste/fulltext, https://globalelectronicscouncil.org/.
39. Director/Head of UNU-SCYCLE/UNITAR BONN. Conference: https://www.ewaste-expo.com/

> fastest-growing domestic waste stream, fueled mainly by higher consumption rates of electric and electronic equipment, short life cycles, and few options for repair.

You can say that government awakened to e-waste under Clinton/Gore and turned its attention away under Bush/Cheney but the question is whether *we* wake up again. There is an e-waste industry, and the market and some government action has done *something*! But think of what we really ought to have.

EXERCISE

Decide on an issue you care about and sign up to receive alerts of regulatory promulgation, solicited advice, open meetings, and who should receive comments on needed legislation. Understand the basic strategy and functioning of the law and think about how it might work differently for better results. Host a meeting at the local library—or convince someone appropriate to host it—to share ideas. Foster constructive public conversation on the problem and options for solving it optimally. Gather perspectives and seek the goal of preserving common and individual rights.

Chapter Ten

Transforming Systems

When you take the long view, you see the need for change in transportation, shelter, food, production, energy, and the economic system that underlies it all. The best example: poisons that are in our bodies at trace levels. You can argue that we are not now seriously affected by trace levels (though we are, in far too many cases, we now know), but that ignores that their presence is new and growing and life can't sustain that trend. That thousands of artificial chemicals are now in the bodies of living things is not good, period. Even if negligible, now we can see where things are headed. The risk assessment that says it's nothing to lose sleep over is inadequate. Common sense tells you we need the risk prevention that asks how it is that this is happening, how can the trend be turned around, and doesn't quit asking.

The facts are giving us imperatives. Our own nature gives us challenges. The facts provide a universal moral concept, that *we must be on the path of reducing these insults, of actively redesigning our systems to reverse the noxious trends, addressing problems at the source together*. The challenge is that we can't do overall change overnight.

While we envision end points, we want our steps to be free at all times to take advantage of new knowledge, recognitions, coalesced will, and opportunities of all kinds. Seeking efficiency and justice are goals that involve all who have a stake, even those who cannot speak.

This is not an impossible job. It's an effort that can take many forms, and a culture of envisioning, of group creativity about common problems can be a powerful flowering. We can use government to foster it. We could, in many places, join ongoing discussions.[1]

1. Just a few examples: a sustainability committee in your town, the planning board, legislative hearings, Food Policy Councils, https://scholars.unh.edu/nren_facpub/143/, zero-waste and green teams, conservation commissions, climate action networks, and regional emergency planning committees.

THE COURAGE TO ENVISION NEW ECONOMIC SYSTEMS

Discussing changes to an economic system can be frightening. A recent Cold War induced the fear of nuclear death over a disagreement about economic policies. Environmentalists have experienced a history of being opposed because they are made out to represent a catastrophic change in our economy. When the idea of allowing the environmentally concerned to be in charge is heard by too many as a recipe for starving or losing your stuff, that has to be faced. Seized as we are by simple stories, and habituated now to sudden escalation of difference, we need to break out of the cultural trap with new stories.

There's an old joke: "What's the difference between capitalism and communism? Well, in capitalism, man exploits man, but in communism, it's the other way around." The point is that we are free to look back on how neither system was divinely inspired, so perfect that all we need to do is put the system in place. What a free people can do is learn from the experiences of societies and keep working to craft the system that works best. That's what our founders gave us—the example of a people forming a more perfect union.[2]

So, what was good in communism? The idea that everyone has a place and a job and is provided for. What was bad? In Soviet Russia and China, the party ran things as dictators. Freedoms were severely curtailed for many and completely for those deemed enemies of the system (millions were killed). What's good about capitalism? It preserves freedom. With that has come prosperity for many. But not for nearly enough, and when the moneyed feel threatened by competition, they can use their power to clamp down and we essentially end up with oligarchies and monopolies. Maybe you are free to move around, but if you don't have money for food, or can't do business because they can so easily undersell you, you're not very free. FDR's Four Freedoms were of religion and speech, and from fear and want. His idea was for the evolution of a more humane capitalism.

So what do we do? Double down our loyalty to our failed ideologies? Continue to pretend that we have to choose between extremes? So many are doing that, but some see that we are free to do something else. James Madison made a thorough study of everything he and Thomas Jefferson could find that was written about governments before, and opened the convention that wrote our Constitution with a lecture summing up what he had learned. The others brought to the table far more extensive knowledge about theory and practice in ancient societies than what today influences our political discourse.

---

2. "Educated Americans of the eighteenth century were as well versed in history and political theory as in the classics." *The Bill of Rights, Its Origin and Meaning*, Irving Brant, 1965, p. 38. See also *John Marshall, Definer of a Nation*, Jean Edward Smith, (1996) for examples of the study of history and thought that provided the foundation for the ability to produce the US Constitution.

We can envision a society in which people have the freedom to begin businesses, and no one can corner so much of the market that they can drive you out. Think of a government that creates a space for a market, and you can apply to sell something there, and if you make something people want, you make a good living. You can even expand and get rich. This is a government-supported market. You would not be allowed to buy up all the space or make it hard for others to enter the market. The government ownership ensures the continuity of the free enterprise idea. The government also makes sure that whatever you are doing doesn't harm others. No big piles of waste that you just leave. No raucous stinky pollution, no poisonous activity, no fraud. The marketplace is clean and safe and equitable and good for everyone. I think of the farmers' markets that cities set up. They don't evolve to be one big farm having all the booths, they provide incomes for many family businesses.

Now marry that beautiful capitalist concept—in which competition produces excellence and continuous innovation incentivized by profit gives society an ever-growing flowering of ingenuity—with the communist or socialist idea that everyone gets a place, everyone has a job. If you can't start a business, have no idea what you can sell or just don't want to engage in production or commerce, the government finds something useful for you to do. In a society that prizes free enterprise, that can even be activities the market is not providing, or providing sufficiently, or where some competition from the public sector will improve the private.

What's stopping us from trying to develop this middle ground? This is the question to ask. It's not a choice between one or the other extreme. It's a choice to us as to whether we will even try to find the middle.

Could it be fear that if we concede there is anything good about communism, that that would be losing a war? Or that we must not concede there is anything good about capitalism? Or is it that we are simply not accustomed to thinking outside of our box? Maybe we can remember that we didn't put ourselves in the box—we were born in it, but now we are grown up, and can step out, and think for ourselves. We do have the ability to have discussion about what a good economic system would be.

That discussion should, of course, consider what could go wrong. People suggest that we limit the amount of profit someone can make or keep, and others then fear that no one would be motivated anymore to create wonderful things. But people create wonderful things because they want to create wonderful things. Teachers, scientists, artists, singers, creators of nonprofit organizations—they all want to make enough to have a good life, but they don't seem to want to be aiming for having billions. Profit is good. Profit as the greatest good is bad.

The fearful also point to the slippery slope, and say don't go down that road, because where would it stop? If you take a dollar from a billionaire, the next thing we know your taxes will go up more than you can handle. These are fears that can't be addressed by being dismissed. It helps to realize that being able to tax billionaires more actually reduces, not increases, the chances that the ordinary person will have to pay more.

It is as if we are not capable of inventing or using brakes. It's only a slippery slope if we pay no attention to that task. We ought to be able to free ourselves from the trap of thinking we must stick with the current system.

LIGHTING THE WAY

What else could go wrong? It's tempting to think that if we take care of people, they will live long lives and have many children and the world will fill up even more. But history shows us that if you take care of people, they don't feel they have to have many children, to care for them in their old age, or to work in the family business or on the farm—to be free help and social security. When women are free and empowered, have choice and are educated, they choose not to have so many children. Population rates decline. This then produces panicked articles about how there's a problem when there's a small workforce and a large retiree population.[3] This is a problem, but it's a temporary one. The world depends on us getting down to fewer billions than we now are, and history amply shows how we can do it.

In order to transform the economic system, we can establish tax benefits for activities that have social benefits, and tax bads, such as pollution, more effectively than we now do. We have tax benefits for nonprofits, why not for cooperatives? Why not for businesses with a social purpose? Why not for businesses that limit the profit that owners and managers can receive, and which distribute their revenues more fairly to workers and the community?[4] Contrast this example, of cultivating the growth of a new economy, with the idea of a law that suddenly changes our existing one, say by placing an

3. For example, a *New York Times* piece expressing alarm about "Our Future with Fewer Births" by Melissa Karney and Phillip Levine, *Sunday Review*, March 7, 2021, prompted a letter from Philip Warburg, senior fellow at Boston University's Institute for Sustainable Energy (March 16), who called it a "false alarm" and wrote: "What America needs is not more people, but more people gainfully employed in high-productivity jobs that are well matched to the overwhelming challenge facing us: ensuring a high quality of life for all segments of American society while reducing to zero the net carbon emissions of our homes, our transportation systems, our offices and retail establishments, our agricultural sector, and our industries." Such a focus could "salvage this gravely battered planet."

4. I call this the Limited and Mutual Profit Entity, or LAMPE, to light the way to a more just economy. Twenty years ago, I attempted to garner interest in one that would share information about the environmental performance of products and companies, called the Environmental Buyers' Information Association. People were interested in the EBIA but one told me he did not like my provision that investors would be limited to three times their initial investment (with excess going to general membership, which would have a vote on whether it should be invested).

unexpected limit on profits. In this model, we use the government to create an innovative entity that operates in the marketplace, and if it succeeds, it simply replaces the predatory, exploitative corporation. Workers lose their chains, and nobody's hurt. The only loss could be the opportunity to "make a killing" in fast stock trading. CEOs seeing this movement building may quietly move on, if it has enough public support. You can scoff and say it's a dream, but it's also a plan. It helps to have one, as long as you are ready to keep revising it as necessary.

We can envision the kind of organizations that we ought to have, and then structure incentives for them to grow, and gently replace the kinds of organizations we ought not to have (for example, those who value profit without sufficient regard for impacts or equitable sharing of benefits). If we think of government as the great nurturer of good things, as caretaker, we can use it to cultivate that which is of benefit, not just to weed out that which is noxious.[5] Our plan can be for a garden in which we can feel blessed to exist. This is preferable to a laissez-faire economics that has no plan at all for making the world better, only a recipe for allowing dominance by a few powerful entities, with no regard to whether they merit their position and with the myth that we benefit from their domination. The gardener is a better primary role for government than the policeman, who should be adjunct.

It is commonly said that the government should not pick winners but that the market should, and while that is a principle of free enterprise, the government must put some limits on enterprise to curb the destruction of the environment and protect public health, so the question is actually not whether we must do this but only, at this point, how. Our attention must be on *how to artfully influence the evolution of commerce while preserving the maximum of freedom within the necessity of constraint.* The government cannot and must not predict who will win or have a chance to win customers in the marketplace—that would likely degenerate into "crony capitalism," but can and must influence marketplace behavior, such as by picking categories or aspects of winners and letting competition take place within the context imposed by our ecological emergency and the social requirement of greater equity.

5. In my own case, I applied for and received about a dozen grants from EPA, which allowed me to: pay someone to investigate reducing chemical use for air quality in schools; create a team to assist state agencies in addressing environmental problems with a cross-media, interdisciplinary approach; demonstrate how toxics use reduction saves money; marry pollution prevention and emergency preparation to demonstrate a new hazard prevention paradigm; and the lead education, Native American environmental and nanotechnology efforts described herein. Those grant opportunities dried up as a result of the brutal cuts to environmental protection, as well as contracts, affecting whole sectors of the economy. Government contracts, grants and cooperative agreements should be stable and substantial as they are key to creating the professionals with expertise in solving our problems.

The effort can lean heavily on incentives. As any dog owner knows, you have to show who's in command or the dog will very likely take advantage of your timidity. There is a prevailing feeling in the world that corporations are in charge. This is not the legal truth, nor need it be the practical truth. Corporations were licensed to operate, with limited liability and rights in many cases identical to people, but we have not yet evolved ways to ensure they behave with needed limits. If a single person went out one night and dynamited a local mountain, that person would likely see a prison term, but a corporation can do it, and dump the broken rock and dirt into a nearby stream. We as a society need to apply legitimate social expectations to organizations as well as to individuals.

When voters act to put incentives in place for companies that do not pollute and share revenues equitably, they are not picking individual companies as winners but using the vote to establish a responsible economic system. They are tilting the odds in favor of good companies and against the rapacious, in favor of the responsible over the sociopathic. This is good evolutionary strategy.

Lester Ward was a founder of the American Sociological Association, who argued against the laissez-faire social Darwinists of his time, the late 1800s. He wrote that:

> The fundamental principle of biology is natural selection, that of sociology is artificial selection. The survival of the fittest is simply the survival of the strong, which implies and would better be called the destruction of the weak. If nature progresses through the destruction of the weak, man progresses through the protection of the weak.
>
> If there is no cosmic purpose, there is at least human purpose, which has already given man a special place in nature and may yet, if he wills it, give organization and direction to this social life. Purposeful activity must henceforth be recognized as a proper function not only of the individual but of a whole society.[6]

It is also important to understand that our present system already tilts the odds in favor of existing companies with significant economic power, by allowing them to participate in the political system. This has led to the stocking of legislative chambers with advocates for powerful companies, and changes in rules that allow them to subvert normal competitive pressures. Political participation by powerful moneyed interests makes the market unfree and corrupts the process and quality of public decision-making.

For example, because of support from the fossil fuel industry rules to control emissions have been prevented, exemptions from environmental laws

6. Richard Hofstadter, *Social Darwinism in American Thought* (1955), chapter 4, 79–81.

have been put in place,[7] and enormous subsidies have been granted. There is already government interference in the market, in favor of dominant market forces. The idea that the marketplace is freely self-determining development is a myth that benefits today's winners.

It is possible to shape the economy so that safer, cleaner, circular economies evolve, that fit within the ecosystem. It is possible to find ways to guide the transition so that it is smooth and easy for people, while achieving the maximum speed toward the needed state of sustainability. But it takes the joint imaginations of many and their combined will to change the corrupted system we have for one that reflects common interests, and because it will involve a decline in exploitative operations and thus affect expected profits, there will be resistance and attempts to disrupt the moment toward widespread consensus. But if we keep in mind preserving to the maximum extent the concepts of free entry, real competition, and minimal government interference within the broad parameters of responsibility necessitated by environmental and public health emergencies, and help people understand that it is only through responsible and humane economics that we can achieve any form of lasting or shared prosperity, perhaps the absurdity of the old binary "capitalism or communism" formula—a distraction as well as distortion—can become apparent, and a richer possibility for political growth grasped.

GENTLER TRANSITION

In 2016, I started working with Quiet Communities, a new organization devoted to reducing noise, to help with the effort to reduce use of gas-powered leaf blowers. These awful things harm workers and the environment as well as making horrible noise. A nearby town, Newton, Massachusetts, was involved in a very contentious fight over a proposed ban. Other places were fighting over such local ordinances. I arranged for the director of Quiet

7. "Oil and gas operations benefit from a range of exemptions or limitations in regulatory coverage . . . the 'Halliburton loophole,' nicknamed for the largest oil and gas services company in the country and passed as part of the Energy Policy Act of 2005 (while former Halliburton CEO Dick Cheney was vice president). Unless diesel is used in the fracking fluid, it exempts hydraulic fracturing from regulation under the Underground Injection Control Program of the Safe Drinking Water Act, the law protecting our drinking water from pollutants. The oil and gas industry enjoys an exemption for certain exploration and production wastes from regulation as 'hazardous wastes' under the Resource Conservation and Recovery Act. Such oil and gas exploration and production wastes could include used fracking fluids, produced water, and many other types of waste. The industry also enjoys a loophole in the Clean Air Act that exempts oil and gas wells, compressor stations, and pump stations from aggregation as major sources that would otherwise have to implement pollution controls once emissions hit a certain threshold. As to the Clean Water Act, Congress exempted stormwater runoff from oil and gas exploration, production, processing, or treatment operations or transmission facilities from certain permitting requirements, provided that such stormwater is not contaminated." Natural Resources Defense Council, https://www.nrdc.org/stories/fracking-101.

Communities, Jamie Banks, to present to the state's environmentally preferable purchasing program, and we ended up creating the first specifications for contracts with government agencies.[8] In my town, the fact that the town maintenance staff have been able to transition away from using gas-powered leaf blowers has demonstrated how it is possible for private contractors to do the same—over time. The sudden ban is a crude tool—there are others that are slower but work better—the town has instituted a "phase-out" over three years, the expected useful life of a gas-powered blower. Some towns with bans have trouble enforcing them. The transition of our lifestyles is best done thoughtfully and gently, however firmly we must press for progress.

THE ENERGY SYSTEM

We know now, finally, we hope with some finality, that we cannot, through our energy use, pour carbon dioxide into the atmosphere, nor release methane as we do. Less discussed is that we shouldn't be pulling toxins from the land of the dead, to roam this surface of the living. We need not be souring our air, oceans, and land. We need not warp the climate. We have the technology, but we are saddled with old habits. Too large looms the idea that the best government does little or nothing when we can be using government to grow clean energy. That is now not just our right—to use government to transform our energy use from dirty to clean—it is our responsibility. Future people will justly consider us deaf and blind to reality or morality if we don't take on this task.

Most especially does this seem a priority when you review the support for dirty energy, how that has even extended to commitments in foreign wars. The United States appears as a Great Satan to some in Iran because it worked with Britain to depose their elected leader, so that the price of oil could be stabilized. US customers who are not mindful of this history may simply regard the depiction of our country as insulting and wrong. Those who are may be impatient for a quick transition of our nation to a more personally related member of the world civilization.

There ought to be a rule against sudden transitions—suddenly changing the rules of a game. But there is no rule against evolving new systems, new expectations of each other. Think of turning a car, you don't wrench the wheel in another direction.

If we ramp down the subsidies for dirty energy, the market would work to bring about clean energy faster, because it would no longer be distorted. It's not about whether we should have central planning or not, because we have

8. Thanks go to Dan Mabe of the American Green Zone Alliance.

it by default. The domination by powerful elites is allowed by a gap in public interest. We ironically end up with a kind of central planning not dissimilar to that imposed by totalitarian or authoritarian states, when government only listens to the powerful or to elites.[9] Rather, we can call on our government to host the kind that builds consensus out of many points of view. Maybe we should call what we need *coordination* and not planning. Planning implies a complete takeover. The coordination should be of visions, so that they coalesce, so that a plan emerges from the process.

One vision that is widely and rapidly forming is the idea of electrification. We can read about this and be glad that people are on it, or we can get involved ourselves in some way to help make it happen and happen right. The path this takes will depend on who's involved.

If we can electrify cars and have heat pumps in homes, we can do a lot to reduce greenhouse gas emissions. Those cars have batteries in them, which can also be used to help store electricity, adding to the resilience and efficiency of the system, and those homes can have solar panels on them. But we can pursue this vision crudely or in a thoughtful fashion. Is this vision anticipatory? Does it ensure that the batteries will be safely recycled? Does it distribute its benefits equitably? Does it cause unnecessary disruption? Does it allow for experimentation? Who will build it out? Who will be able to participate in it?

We can shape the path for this new technology so that many people are part of the build-out and get the benefits, and no one is hurt, or we can allow the sharks to move in and capture the new technologies, and charge rent to the rest of us forever to use them. The grasp of a society that chooses the former over the latter is through the political will that comes from a coalescing vision of how the system can and should be changed, one that is faster than the purely commercial interests, faster than their ability to lock up proceeds. The vision of electrification teeters now between the developing grasp of the coalescing public interest and the crowd of grasping hands from well-capitalized entities.

What we have now is a system that was designed for large centralized power plants that send electricity in one direction, out to customers, and most of the system was built by investors who want a return on their money. This has led to the situation where many utilities are actually resisting the

9. Vice President Dick Cheney's 2001 Energy Task Force meetings apparently violated the Federal Advisory Committee Act by enabling participation by oil company executives and lobbyists and excluding environmentalists, keeping its deliberations from the public. The administration's policies strongly supported conventional, climate-damaging energy, https://sourcewatch.org/index.php?title=Cheney_Energy_Task_Force.

changeover to a clean energy system, because it conflicts with their business plan.[10] If they have to pay all those homes with solar panels for the energy they put in, that's a big reduction in their profits as things are now set up. The University of California's Leah Stokes points out that while some believe Texas has the "greatest potential for solar in the country," it is instead a "below average" clean energy state, and we see this in many states:

> Since 2011, opponents have blocked clean energy laws from being passed in new places, leaving many states without any target. Others have begun to roll-back or weaken clean energy policies. When these strategies have failed, electric utilities and other polluting companies made the fight bigger, using the public, the parties, and the courts to short-circuit progress on climate policy.[11]

We can recall that we have public utility commissions that are supposed to regulate utilities in the public interest, and that the idea of the system was that they were allowed a reasonable profit for providing us with electricity. Now the public interest is for us to have a different kind of system.

The ideal system would have them building out the grid so that it can take power from many sources, and travel in both directions, and would allow many more people to make money from providing electricity. That is a distributed system rather than concentrated. The ideal system would also be resilient, with microgrids all over the place, that would provide people with stored energy from a local source when the power goes down on the greater grid. Because it would have so many local sources, losses from sending electricity over a distance would be reduced. Local sources could mean local expenditures become local revenue, a recipe for building and preserving community. This is a different model than what works for the investors' business plans. It is more efficient and more just. What is before us is to envision the better system and then plan out how to use the powers of the public utility commissions, and the federal regulatory system over them, to foster this evolution. We should be able to come up with a fair deal if we involve communities in this energy future planning. But in many places, this is not happening, the political system is not moving to address the public interest in the new system. The investors want to hold onto the deal they're used to, and not enough of the general public understands that there is a choice to be made here, not enough even have anything like this vision of how communities as

10. For example, see https://www.audubon.org/news/why-electric-utilities-are-resorting-dark-money-and-bribes-resist-renewables. Some also see it as a business opportunity, https://www.utilitydive.com/news/how-4-utilities-are-turning-distributed-generation-into-an-opportunity/300902/. But the question is: Will the new system be one that provides benefits to investors or to the public? Will we get the balance right?

11. *Short-Circuiting Policy: Interest Groups and the Battle Over Clean Energy and Climate Policy in the American States* (2020): 2–3.

well as the entire planet could benefit from a new system. The acceptance of a private company providing the public good is deep, because that is our history in large part. Investor-owned utilities supply most of our electricity. The logic of investment is for future return in the form of dollars going back to the investors, not the community. Government intervention is necessary to counter this understandable logic enough to produce the right balance of interests.

Gerry Braun, Chair of the Integrated Renewable Energy Systems Network, notes local communities need the capacity to participate in how the new system is set up:

> half of all decarbonization opportunities in the US are locally actionable. So are energy resilience opportunities. Climate-driven disasters in many US states continue to disrupt local energy supplies and local economies. Urgency for both energy resilience and decarbonization is greatly amplified in areas recovering from recent climate related disasters. Local public investment options may be limited in these areas and many other cases, creating a need for local planning that carefully quantifies and evaluates costs, benefits, and risks.[12]

Braun argues for a balanced approach that makes use of the centralized grid system but puts greater needed emphasis on local development, to aid local economies, increase resilience, and speed our build-out of renewable energies.

His perspective encompasses energy sources often left out of the discussions that focus on electrification only, noting that renewable biomethane, developed from sources such as sewage or agricultural waste, can substitute for nonrenewable geologic methane. A recent article in *Renewable Energy*[13] finds that while initially costly to develop, the potential for connecting sewage-plant biogas to existing gas delivery systems is quite feasible and will benefit the infrastructure, local economies, and climate goals. While fossil fuel and nuclear advocates often claim we need an all-of-the-above strategy to fulfil our energy needs, we don't need highly polluting sources, and there are many clean energy choices to consider in addition to solar and wind, which can be developed further. While some focus on the need for more transmission lines so that wind power or solar farms can be linked to overcome the problem of intermittency (if the wind is not blowing where you are, it might be blowing elsewhere), others focus on the need for battery and other storage to help balance capacity and demand. A balanced build-out requires balancing the inputs and outputs on an instantaneous basis, and the interests

---

12. https://www.iresn.org/news/2020/10/14/local-energy-transition-strategies.

13. Stürmer, Novakovits, Luidolt, and Zweilerb, "Potential of Renewable Methane by Anaerobic Digestion from Existing Plant Stock—An Economic Reflection of an Austrian Region," *Renewable Energy* 130 (January 2019): 920–29.

of all concerned, including the voiceless future generations who need us to sort this out, on a long-term basis.

The grid needs modernizing, as we have "congestion problems" that cause the utilities to put "caps" on what local sources can feed in. Besides the need for equipment designed for a system drawing power from many sources, we need to reform how we charge for use and pay for feeding in energy. Time of use and locational pricing can help match the incentives to the need, so that the market works more like the central efficient planner (the invisible hand) we like to think it can be. Energy policy is very complex, but some things are simple, such as that the closer we can put generation to use, the fewer transmission losses we see, and the more we know about the cost of use, the more we will adjust our use to reduce those costs.[14] More "cost-reflective" charges can speed the adoption of rooftop solar and household heat pumps.[15] Incorporating the social cost of carbon emissions into planning is a similar idea—a simple concept that embraces both the public interest in seeing carbon emissions decrease and letting the market work. Community choice aggregation lets ordinary people bargain—through the medium of their local government—for bulk prices, like any large corporation, and provides the basis for fostering the growth of clean energy to supply contracts directly made by communities with aggregated purchasing power. With CCA, communities can plan a clean energy future. But most communities need help taking advantage of such opportunities and to help them have a voice in the proceedings that determine how the energy system will grow.

Electrification and use of greener biofuels (greener than that made from corn, for example) for vehicles and home heating will help a great deal, but we still have much of heavy industry that needs high density power to do its job. How will they replace the coal-, oil-, or gas-powered furnaces they need for high heat and pressure? Cleaner production will help, using the green chemistry principle that you want to mimic nature, which creates many things at room temperature, finding ways to bring things together that belong together, as opposed to smashing and burning.[16] Until that good science takes over, those who need that high level of power can use hydrogen. When you burn hydrogen, the emission is water. The issue is how you get the hydrogen in the first place. Often, hydrogen is made by processing natural gas, which

---

14. See, for example, Josh Smith's "The Electrical Grid is Changing" (April 5, 2019), in *The Benchmark*, a publication by the Center for Growth and Opportunity at Utah State University, https://medium.com/cgo-benchmark/the-electrical-grid-is-changing-6a2e89b04a.

15. Spiller et al., "The Role of Electricity Tariff Design in Distributed Energy Resource Deployment,"

Environmental Defense Fund Economics Discussion Paper Series, EDF EDP 20-03, posted December 8, 2020, on SSRN, https://papers.ssrn.com/sol3/papers.cfm?abstract_id=3711073.

16. Paul Anastas and John Warner, *The Twelve Principles of Green Chemistry*, https://www.acs.org/content/acs/en/greenchemistry/principles/12-principles-of-green-chemistry.html.

has polluting consequence. But the technology for producing hydrogen using electricity, which you can get from solar, is advancing.

How to make all this happen? Subsidies, expediting permits, funding research that demonstrates feasibility. Convene meetings to discuss how to speed development. Find the roadblocks and remove them. The many tools of government can be examined and applied. Government spurred the creation of moonshot technology, and the internet, which were not preexisting.

Our plans can promote free enterprise as well as community interests, local as well as centralized clean energy production. It's not a binary choice between parties but a choice of actions and policies to best balance everything. This will not happen when many localities do not have staff or money to invest. This will not happen when state utility commissions primarily consult with privately owned utilities to develop the system we all need, and the public is not even aware of the decision-making, nor does it have help to parse the questions. It might happen if enough people took enough of an interest and got involved.

In a 2013 article, University of Minnesota Law School professor Hari Osofsky and Florida State University School of law professor Hannah Wiseman described a complex and confusing energy regulatory system that poses many challenges:

> inadequate regulatory authority, simultaneous overlap and fragmentation of regulation and institutions, and the difficulties of including key public and private stakeholders while avoiding inappropriate regulatory capture, such as when powerful utilities or oil companies gain control of regulatory processes to protect their private interests at the expense of the public.

They propose a "dynamic federalism" approach that would be "responsive to these governance challenges through (1) creating needed authority; (2) reducing fragmentation; and (3) allowing for high levels of involvement from key public and private stakeholders that accommodates meaningful input without capture."[17]

In the news in the winter of 2021 was the calamity in Texas in which a sudden deep freeze froze operations at energy plants and people lost power, pipes burst, and communities were left in the freezing dark and without water. Then, some consumers received outrageously high electricity bills. While Governor Greg Abbott blamed wind turbines, commentators pointed out that wind is a minor part of the infrastructure and that conventional fossil fuel energy plants had been warned that such a deep freeze was possible but had not adopted the measures commonly used in more northern areas. Texas had purposely kept

17. https://scholarship.law.umn.edu/cgi/viewcontent.cgi?article=1190&context=faculty_articles.

its grid within state borders in order to avoid federal control, an arrangement stemming from the New Deal, when monopolies and stock manipulations had created economic havoc and the Federal Power Act brought *interstate* sales of electricity under regulation. Texas's governance-avoidance policies allowed companies to take short-term perspectives and gamble that it would never get so cold that they needed to invest in preparations. Their own fossil-fuel use has exacerbated the climate changes that caused the conditions for which they were not prepared. The lack of preparation suggests the owners are more concerned with speculative uses of money than meeting their responsibilities. The sudden high bills reflect contractual arrangements unfairly placing too much of the risk of speculation on consumers, rendering the entire arrangement one that cannot be seen as meeting the public interest. The argument of Samuel Insull was that a utility is a natural monopoly,[18] and the social contract that developed out of abuses is that we allow the monopoly in return for them being part of a public utility system. It's not laissez-faire.

Ceasing drilling and coal mining on federal lands will at the same time help conserve wildlife and the natural beauty that is a balm to our spirits. There will be costs, and lost jobs, and we can act to ease the transition for those wedded to the old ways. They are not our enemy. As Pogo said,[19] we have met the enemy, and he is *us*. We can find the emotional energy to be our friend in the process.

A few years ago, a young man came to a meeting of my town's sustainability committee to pitch the services of his company, which installed insulation and improved the energy efficiency of houses. He said he wanted to Kick Oil. I said that that might sound fun but it would likely not sound right to older people, who were not at war with the industry, who accepted it as part of their lives. Many of them would still be heating their homes with oil as they reduce their heat losses with his company's insulation. I then apologized for advising him on his pitch but he was appreciative and said he would modify it. My impression was that he was excited by the idea of challenging Big Oil, that he liked the idea of giving them a little jab, getting them back a bit. I am not sure that I didn't spoil some of my colleagues' fun as well. But kicking can inspire kicking back, and we're talking about powerful interests. A less simplistic view of our situation might help inspire a more creative and effective response. A gentle but firm kick may be appropriate but it seems that's the easy thing to do. There is real work in front of us: to generate the positive vision of what replaces our existing system, so that we have the political ability to create and sustain coordinated movement.

---

18. https://www.chicagotribune.com/news/ct-xpm-1994-11-27-9411270177-story.html and https://www.britannica.com/biography/Samuel-Insull.

19. Walt Kelly's cartoon possum, https://library.osu.edu/site/40stories/2020/01/05/we-have-met-the-enemy/.

LAND AND FOOD

In the United States, plantation agriculture competed for land with the kind that is honored in cultural memory, where a family raises its own food and sells some of its produce in markets to raise a little cash. Plantation agriculture has been based on the exploitation of people and land, and it exhausts the soil as well as the people. It is not pursued for subsistence but for profit. Its beginnings can be seen in the enclosure movement in England, by which lords of the land took common property for their own with the rationale that they would make it more productive.

John Locke, one of the great inspirations of the founders of the American democracy,[20] asserted in his *Second Treatise on Government* that the world was given "to men in common" by God. But God also gave us the right to property.

> Whatsoever then he removes out of the state that nature hath provided, and left it in, he hath mixed his labour with, and joined to it something that is his own, and thereby makes it his property. It being by him removed from the common state nature hath placed it in, it hath by this labour something annexed to it, that excludes the common right of other men: for this labour being the unquestionable property of the labourer, no man but he can have a right to what that is once joined to, at least where there is enough, and as good, left in common for others.[21]

At least where there is enough. Never mind that the labor theory of value seemed to overlook, in practice, whether the labor was provided by slaves or poorly paid. If you focus on the idea that the earth *should* be made productive, then private property can come to be more important than common possession. Unimproved land is wasted. The prime directive is production, make something you can sell. It is the directive you would expect a servant receives.

What's valued is what we take from nature and make something out of and to do this is a duty and a benefit to all. Locke was part of the inner circle of the Earl of Shaftesbury. This philosophy supported the perspective of landlords. But note that it requires the condition:

> *Where there is enough, and as good, left in common for others.*

---

20. See, however, Thomas Ricks's 2020 *First Principles: What America's Founders Learned from the Greeks and Romans and How That Shaped Our Country*, which stresses the greater importance of defenders of the Roman Republic (such as Cato or Cicero).

21. Section 27, https://resources.saylor.org/wwwresources/archived/site/wp-content/uploads/2012/09/POLSC2012.3.pdf.

The movement to privatize common lands for increased production evolved over hundreds of years. Farmer and coeditor of the *Ecologist* Simon Fairlie writes:

> as medieval England progressed to modernity, the open field system and the communal pastures came under attack from wealthy landowners who wanted to privatize their use. The first onslaught, during the fourteenth to seventeenth centuries, came from landowners who converted arable land over to sheep, with legal support from the Statute of Merton of 1235. Villages were depopulated and several hundred seem to have disappeared. The peasantry responded with a series of ill-fated revolts. In the 1381 Peasants' Revolt, enclosure was an issue, albeit not the main one. In Jack Cade's rebellion of 1450 land rights were a prominent demand. By the time of Kett's rebellion of 1549 enclosure was a main issue, as it was in the Captain Pouch revolts of 1604–1607 when the terms "leveller" and "digger" appeared, referring to those who levelled the ditches and fences erected by enclosers.[22]

"Britain set out, more or less deliberately, to become a highly urbanized economy with a large urban proletariat dispossessed from the countryside, highly concentrated landownership, and farms far larger than any other country in Europe." Locke, however, notes that the "appropriation" of the land, to improve it, worked no prejudice to any man,

> since there was still enough, and as good left; and more than the yet unprovided could use. So that, in effect, there was never the less left for others because of his enclosure for himself: for he that leaves as much as another can make use of, does as good as take nothing at all. No body could think himself injured by the drinking of another man, though he took a good draught, who had a whole river of the same water left him to quench his thirst: and the case of land and water, where there is enough of both, is perfectly the same.[23]

In a world of infinite resources, perhaps this could be true, if it also ignored the issue of equal access to opportunities. Those who consider private property sacred, primary over common rights, should read more carefully the original patron of that idea. There is not a "whole river of the same water left," nor of land. Fairlie writes that the result of the evolution he describes is that "nearly half the country is owned by 40,000 land millionaires, or 0.06 percent of the population."

In *Changes in the Land*, ecological historian William Cronon contrasts Henry David Thoreau mourning the loss of wildlife with the historian Edward

---

22. "A Short History of Enclosure in Britain," *The Land* (Summer 2009): 20, https://www.thelandmagazine.org.uk/sites/default/files/enclosure%20low%20res.pdf.
23. *Second Treatise*, Section 31.

Johnson, who in 1653 praised the transformation of the landscape into "a second England for fertilness." Cronon observes: "In this vision, the transformation of wilderness betokened the planting of a garden, not the fall from one; any change in the New England environment was divinely ordained and wholly positive."[24]

Jedidiah Purdy of Duke Law School uses the term "Providential Republicanism" to describe how the Europeans thought about the new American continent, the view that:

> the natural world is made for productive use: nature has a telos, to be fruitful and support human life, and this purpose is realized only through human labor. Moreover, small-scale private ownership of land, the sort associated with the republican thought of the eighteenth and nineteenth centuries, uniquely produces fruitful use, so that political freedom and economic productivity go hand-in-hand. The dignity of the smallholder or pioneer and the progress of a continental "empire of liberty" go hand-in-hand. Much of the continent's private property in land was created under the banner of providential republicanism, and its tropes persist in property-rights movements, calls for local land-use sovereignty, and resistance to regulation as a centralizing, tyrannical, and possibly European scheme against liberty.[25]

Policies that have allowed the concentration of land use in the hands of a few seem to have overshadowed this idea of most fruitful use through widespread small-scale private ownership.

Purdy cites President James Buchanan's Second Message to Congress:

> Immediately upon the formation of a new Territory people from different States and from foreign countries rush into it for the laudable purpose of improving their condition. Their first duty to themselves is to open and cultivate farms, to construct roads, to establish schools, to erect places of religious worship, and to devote their energies generally to reclaim the wilderness and to lay the foundations of a flourishing and prosperous commonwealth.

Government existed to support this, and when the army marched through the Indian Territory it had

> a powerful effect in restraining the hostile feelings against the United States which existed among the Indians in that region and in securing emigrants to the far West against their depredations.

---

24. *Changes in the Land* (1983), 4–5.

25. "American Natures, the Shape of Conflict in Environmental Law," *Harvard Environmental Law Review* 35 (2012): 169–228, https://scholarship.law.duke.edu/faculty_scholarship/2356/.

The settlers believed their occupation was justified as the work of civilization. The signing of the Homestead Act and the Morrill Act that established land-grant colleges served, from one perspective, the laudable purpose of enabling people to sustain themselves through agriculture, and the establishment of educational facilities. But a recent *High Country News* project asks, where did that land come from?

> Nearly 11 million acres of Indigenous land. Approximately 250 tribes, bands and communities. Over 160 violence-backed treaties and land seizures. Fifty-two universities.[26]

The organization One Green Planet analyzing data from the UN's Food and Agriculture Organization finds that half the world's habitable land is used for agriculture. Seventy-seven percent of that land is for raising livestock.[27] Wenona Hauter points out in *Foodopoly* that only "twenty food corporations produce most of the food eaten by Americans, even organic brands. Four large chains, including Walmart, control more than half of all grocery store sales. One company dominates the organic grocery industry, and one distribution company has the stranglehold on getting organic products into communities around the country."[28] *Four hundred thousand families received lands from the Homestead Act, but the providential republic has been captured.*

The productivity of agriculture boomed when millions of tons of guano were shipped from Peru to Europe and North America, until the bird feces that accumulated for eons was exhausted, and then it boomed again when the Haber-Bosch process that takes nitrogen from the air was invented, and again when Norman Borlaug headed up what was called the Green Revolution, that used new plant varieties coupled with inputs of fertilizer to greatly boost food production. But as former president of Oxfam Ray Offenheiser points out:

> Borlaug was very committed in his early years to working with very poor farmers in Mexico and really trying to create a crop and a product that would improve their livelihoods and reduce malnutrition and generate disposable income for those families. As time went on, however, one of the peculiar effects of this was that because you had to purchase inputs, and because you had to have access to water, to some degree it became the case that in many parts of the world only the more capitalized farmers can actually get access to the money needed to buy the fertilizer, or buy the pesticides or herbicides, and also have access to water.

---

26. *Land-Grab Universities*, Robert Lee, Tristan Ahtone, Margaret Pearce, Kalen Goodluck, Geoff McGhee, Cody Leff, Katherine Lanpher and Taryn Salinas, https://www.landgrabu.org/.
27. https://www.onegreenplanet.org/news/chart-shows-worlds-land-used/#:~:text=As%20the%20chart%20details,%2071%20percent%20of%20our,for%20grazing%20or%20land%20to%20grow%20animal%20feed.
28. *Foodopoly: The Battle Over the Future of Food and Farming in American* (2012), 3.

> So to some degree you've got consolidation in the agricultural sector and instead of actually necessarily helping small farmers in certain instances, more in Latin America then in Asia, you ended up displacing the small farmers that Borlaug had intended to help. So those were other socioeconomic and political implications to the technology that were not forseen when Borlaug, I think with the best of intentions, was breeding for these new varieties to help these small farmers.[29]

The Norman Borlaug Institute for Plant Science Research recognizes this problem, as well as the "very high environmental impacts" of vast application of chemicals and water use, and is now "committed to developing low input/low environmental impact/high yield/high quality strains of crops," in an effort to "satisfy the need for efficient sustainable agricultural production in both developing and developed countries."[30]

There's a long way to go. We need a "regenerative" agriculture that builds soil, refrains from chemical loading, doesn't crowd out all species besides those we use for food, nurtures those who grow and provide nutrition where it's needed. The International Union for the Conservation of Nature produced a report in 2020, *Common Ground*, to communicate a vision of what's needed, which envisions an overall approach of public policies to help communities maintain biodiverse soils and landscapes, establish "biodiversity-friendly supply chains" and promote responsible finance and investment; and applying agroecological practices, governance and planning, in order to promote "human wellbeing, profitable and resilient farming, healthy diets, food and culture," and provide societal benefits such as economic, climate, food and water security, and social justice.[31]

New approaches have been developed, but they compete with the plans of major industrial agricultural powers, which dominate the discussion. A google search for "sustainable agriculture" turns up pages of information from such corporations as Bayer Crop Science. But if you look long enough, you will find alternative visions, such as that from Common Dreams, which points to the "agroecology" being performed by small farmers, in a "bioregional context."

> Agroecology uses low fossil fuel-based inputs and has a better carbon footprint than industrial agriculture-based practices. The wide array of agricultural practices that are part of agroecology which include permaculture, agro-forestry, organic farming, biodynamic farming and zero budget natural farming, make it

29. https://www.pbs.org/wgbh/americanexperience/features/green-revolution-norman-borlaug-race-to-fight-global-hunger/#:~:text=When%20Norman%20Borlaug%20set%20out%20after%20World%20War,1970%20for%20his%20role%20in%20fighting%20global%20hunger.

30. http://www.agbioworld.org/biotech-info/topics/borlaug/green-revolution.html.

31. https://www.iucn.org/theme/ecosystem-management/our-work/agriculture-and-land-health/common-ground-report#q1.

optimal in soil carbon sequestration and thus a powerful tool to help mitigate climate-related risks.

Small farmers across the world are continuously adapting their way of living and farming, enabling them to withstand the climate breakdown and minimize GHG emissions. After experiencing climate impacts—drought, irregular rains, hailstorm, reduced monsoon and increased pest attacks—small farmers are developing their own strategies to combat the climate crisis. These include:

> water harvesting systems,
> better irrigation techniques that include drip irrigation,
> the use of traditional seeds, bio-fertilisers and bio-pesticides,
> mulching, multiple cropping and mixed cropping practices,
> gathering information on timely weather reports,
> proper planning of agricultural practices,
> biodiversity conservation and
> the increased use of solar energy.[32]

Hauter tells us that "Breaking the foodopoly and fixing the dysfunctional food system require far-reaching legislative and regulatory changes that are part of a larger strategy for restoring our democracy." She argues that we must not be apathetic or cynical, and that history shows the political system can be reformed.[33] One area of focus must be antitrust—for the consolidation of the industry prevents the changes we need. Enforcement responsibility is divided between the Department of Justice, the Federal Trade Commission, and the Department of Agriculture, none of which have taken up the cause. Government subsidies now fail to support farmers, and actually benefit agri-businesses while encouraging overproduction. The Farm Bill's system of subsidies can instead focus on "bringing back strategic grain reserves, requiring that farmers leave a portion of land fallow, and maintaining minimum price floors for crops," which would protect farmers better and prevent soil exhaustion.

> Future Farm Bills should focus on leveling the playing field for independent farmers, ranchers and food processors and redirecting rural development programs to rebuild missing infrastructure that can serve regional food systems, not corporate supply chains.[34]

32. "Sustainable Agriculture in the Face of the Climate Crisis: Small Farmers Leading the Battle in Asia," March 12, 2021, https://www.commondreams.org/views/2021/03/12/sustainable-agriculture-face-climate-crisis-farmers-leading-battle-asia. The author is listed as GRAIN, "a small international non-profit organisation that works to support small farmers and social movements in their struggles for community-controlled and biodiversity-based food systems."

33. Hauter, *Foodopoly*, 287.

34. Ibid., 295.

Gus Schumacher, once Commissioner of Agriculture for Massachusetts and Undersecretary of Agriculture under Clinton, founded Wholesome Wave,[35] a nonprofit devoted to helping low-income populations gain access to fresh food, and was a force behind such initiatives as Veggie Rx, in which doctors wrote prescriptions for children who needed better nutrition, and programs to double the value of food stamps and enabling their use at farmers' markets, which he also worked to foster. I had the great privilege to host him as a speaker at events promoting these ideas, and to work with him to try to get health insurance companies to fund better nutrition when medically advised. He often told a story that he called "the broken pear box." It explained why he devoted himself to such pursuits. It is recounted in a 2011 *Nation* article:

> Schumacher was helping his brother, a farmer outside Boston, load a truck after the Dorchester farmers' market closed when he dropped a box of Bosc pears in the gutter. A woman passing by bent down and retrieved the damaged fruit, explaining that she was on food stamps but still couldn't afford fresh fruits and vegetables for her children. Schumacher ended up giving her an additional ten pounds of apples and pears. "People should not be picking fruit out of a gutter," he says.[36]

As with every other issue discussed herein, there are too many changes that need to be written to adequately cover what matters, but the point of the book is to enter the "solution space" to prompt discussion of options for bettering our situation. Governments can help foster cooperatives for storage and distribution as replacements for companies that extract the monetary values and cheapen food value and make life harder for producers. Governments can help train new farmers and help them get the land they need and to keep it. Governments can help farming transition to methods that regenerate the soil and reserve land for biodiversity, and cease using chemicals that kill insects, birds, and foul waters. Governments can help develop humane livestock practices through training, verification, organizing conferences and cooperatives, providing grants and loans, educating the public.

By focusing on people, from those who grow the food, to those who wish to eat it, we can design a system that makes sense, because people wish to see animals treated humanely, and people wish to protect the environment we use to grow food. An apparent separation between the interest of humanity and the interests of all other living things is illusory. These two sets of concerns are connected at every conceivable point.

We need to call on government to help us together devise many versions of such systems, so we can see what works. We can try to help people who

35. https://www.wholesomewave.org/.

36. Melanie Mason, "Food Stamps for Good Food," *The Nation*, March 28, 2011, https://www.thenation.com/article/archive/food-stamps-good-food/.

have had land taken away from them, or were denied loans available to others, or have been ill-fed, to help themselves and the soils and ecosystems around them, to make their immediate environments enriching and ecologically healthy. It's a different approach than sitting back and waiting for the industrial system to do it for us. It's going to take a little more involvement than that, to retake some of the land that John Locke pointed out was given by God to all. We can place his idea of the labor theory of value, modified by respecting who actually does the labor, within our knowledge of finite limits and replace the consumer/producer co-dependency model with the relationship of the partner, member, fair exchanger, engaged citizen and responsible practitioner.

SHELTER

When we think about supplying our needs sensibly, for example, without harming others or the environment, we envision homes and buildings that have achieved "net-zero" status—that is, they generate themselves all the energy they need or are fully powered by renewables, even if they are somewhere else. We hope that they are also collecting rainwater, perhaps to use for flushing toilets. We picture them using the "graywater" that comes from showers and sinks to water the gardens that surround them, and if they have lawns, that those lawns do not require constant watering or chemical feeding, because they have been enriched with compost, made from the food waste generated within the building or home perhaps, and they are not just grass, but include native vegetation—more like wildflower and nitrogen-fixing clover meadows, with bees, bunnies, butterflies, hummingbirds.

If mowed, it is done with an electrical mower that mulches leaf nutrient right into the soil. They are raked and not attacked violently with blasts of air created by fuel combustion, which blows insect nests to pieces and everything into the air for people to breathe.

We imagine that the home was constructed out of nontoxic and renewable materials, such as adobe, bamboo, compressed earth block, cork, hemp, lumber that has been sustainably harvested and certified as such; natural fibers, stone, recycled metal, linoleum, straw bale, wood; and insulated with cellulose, spun rock or recycled denim, and all the glues are plant-based and don't have formaldehyde, and if there's brick or concrete, it's recycled, and sustainable sourcing and production are reliably verified.[37]

37. The verification and certification processes need consideration. There ought to be several ways of doing it, having only one system is too easily corrupted. The analogue is to think about how the separation of powers, provisions for citizen suits and transparency help ensure that government systems remain faithful to the public interest.

We may have to rely on some materials that don't have that high an ecological and health preservation value, but the point is that we try to incorporate these kinds of materials if they will work. We might have a garden on the roof, or even green growing things on the walls, or we might have incorporated solar panels into the building or placed them on the roof or over the car park, and there could be a wind turbine, but that might cause vibration or noise, so we need very good design to avoid those problems. If we have really gone the distance our bodily wastes are recovered in a composting toilet, flushed to treatment that generates biogas and organic fertilizer or feeds plants and fish and the phosphorous is recovered from the urine to feed back to the soil.

We are using heat pumps to cool and heat the building, with the electricity for the pumps supplied by solar energy, which also feeds our car, which plugs into our house at night and we might use its battery if we need to. Our house is superinsulated but not with toxic foam, and we lose a little of that saved energy because we have a ventilation system to make sure we get lots of fresh air. The system has a heat- and moisture-recovery capability so that we don't lose too much energy and keep our humidity stable. Helping to warm the house or building is a ground-source heat pump that also helps keep humidity the same all year round, adding to our sense of comfort, because it draws from a constant temperature under the ground. Possibly we have a neighborhood geothermal system that feeds heating and cooling air to each house, using some of the former gas delivery system, refurbished. The stabilizing of heat and humidity increases the durability of the building, reducing warping and deterioration.

Our sustainable house or building takes advantage of intelligent siting, the windows allowing sun in the winter and shaded in the summer, and the surrounding trees that protect it from both heat and cold make it beautiful. The water system is efficient, free of toxins, and people take their shoes off before they come in the house so as not to track in anything. No harsh chemicals are used for cleaning, polishing, or fixing things, and nothing is stored that will off-gas and be breathed or start a fire.

We can envision this house; it has been shown to be feasible. Leadership in Energy and Environmental Design (LEED) is the best-known system for green building design, designed to prompt transition, which it has achieved. Its very flexibility, which allowed it to be so successful at changing things, means that some projects can downplay certain aspects of sustainability by being very strong on others.[38]

---

38. Begun in the mid-1990s by the Natural Resources Defense Council's Robert Watson and the US Green Building Council and "launched with the help of the Federal Energy Management Program in August of 1998." An example of government and nonprofit partnership that has fostered more responsible private sector evolution, https://sigearth.com/leed-history/.

A description of a more fully flowered ideal is the Living Building Challenge,[39] which organizes its goals as a seven-petal flower: the Place Petal has reasonable density, invites natural systems "back into the daily fabric of our lives." The Water Petal addresses the "energy and chemicals involved in transporting, purifying and pumping water" and redefines "wastewater" as a "precious nutrient and resource." The Energy Petal sees the places where we live, work and play as "catalysts for a healthy and resilient future." The Health and Happiness Petal ensures that "indoor spaces have healthy air and natural daylight." The Materials Petal seeks an "economy that is non-toxic, ecologically restorative, and transparent." The Equity Petal "envisions a built environment that is welcoming and accessible, and that enables all people to participate, prosper, and reach their full potential." The Beauty Petal embraces "a biophilic design process that emphasizes that people and nature are connected and the connection to place, climate, culture and community are crucial to creating a beautiful building."

Ultra-low-energy housing is the focus of the Passivhaus Institute; toxic-free living is easier with the Pharos database of the Healthy Building Network, which has hazard, use, and exposure information on 135,414 chemicals and 172 different kinds of building products; BuildingGreen can train you on every aspect of sustainable shelter, and you can find architects and designers now by searching for "green" or "sustainable" architecture. You know that the field has achieved an important evolutionary leap when there is development of the art of assessing the value of green buildings, and you can find at the website of the Appraisal Institute information about how this done.[40] At some point, the value of investing in creating homes and buildings will reach a tipping point, and it may be that we are getting close. What is the role for governance, and the citizen who wants to be involved in accelerating the evolution we need?

Building codes are what most builders build to. If the code says you have to have natural plantings, and you need to provide electrical charging capacity for vehicles, and you must not use petroleum derivatives to seal the driveway, then this is what will happen. The statewide code may preempt the local code, so it may be on the agenda to have flexibility so that local standards can be stricter, or to make everyone meet sustainable goals. But with each new requirement comes complaints from builders, and if you were a builder, you too would feel that it is tough enough to plan and execute your job without all those irritating rules, so codes have to be crafted with an eye to what's doable. In addition, we don't want compliance with new building codes to

39. https://living-future.org/lbc/.
40. https://www.appraisalinstitute.org/education/education-resources/green-building-resources/.

be a disincentive to doing needed renovations (because they trigger the new requirements), so we have to construct these drivers carefully.

If it's not doable to seal driveways with natural materials, then governments should be pressed to fund research, perhaps through universities cooperating with building trade associations, to develop less toxic ones that work, and to work with people at all levels on the transition, always remaining in the doable, so that government works *more like a ratchet wrench than a hammer*. With lead times for adjustments, we can fruitfully develop the codes we live by, subjecting them to the tests of reasoned consensus to evolve gently and intelligently.

To stop builders from cutting down trees just so they can more easily move their heavy equipment around the lot, we can pass stronger tree ordinances, or better, something at the state level, to make them pay enough to be a disincentive. The market signal of the buyers who would pay more for a house with trees is not visible to the builder, who wants to move his materials around easily. The trees the builder takes down for convenience of construction, that the next owner may have wanted to remain, are also important to the community.

We can envision planting native bushes and trees, that don't need watering or chemicals, and flowers that help pollinators to thrive, the houses in living ecosystems connected by green corridors in bioregional arrangements. We can envision alternatives to all the hazardous chemicals used and mobile pickup for those who have difficulty getting containers of dangerous wastes to a collection point, coupled with education to prevent further need to generate such wastes in the future.[41]

We all need classes in these things. It should be taught in high school, as preparation for entering the world as adults. It should be offered to adults for free. If you can help people to understand the basics of these issues, they can guard against vendors who just want to sell something. I installed central air conditioning on my upper floor, because the cool air descends, and I wanted to keep my lower floor windows open for fresh air. It works. The man who installed it thought that because I am environmentally concerned, I would want a reverse-osmosis water treatment system. I asked him why I should have one—what contaminant would it treat that I had in my water? He said he had no idea, but that people who care about the environment were buying them and he could sell me one. In fact, they aren't good for the environment when not needed; they use energy and they create a concentrated brine you need to discharge. They are for health, but if you have clean water, you don't need it. I live in a place where the water is pretty clean, except for the

41. Mansfield, Massachusetts, Public Health Director Amy Donovan-Palmer just held the first mobile pick up in the area with the assistance of ACV Enviro, April 24, 2021, with adjacent towns Foxboro and Norton.

possibility of lead in the pipes, and what I do is never take that first draw of water in the morning. I fill my teapot the night before, and if I forget, I use the water on the plants or just let it run till it is cold from the street. That way we avoid any of the lead that it might have picked up sitting in the pipes all night long.

When I was working for the state, there was a big fuss about a sick building, where people were experiencing terrible smells, and the air was indeed toxic, as a coating had failed to cure. I was asked to pull together a conference on sick buildings and what to do about them. After some research I found Joseph Lstiburek of the Building Science Corporation, at that time (circa 1990) the only expert on the topic in the area I could locate. He came and gave an excellent presentation on ventilation, and at one point either lit a cigarette or used a little puff bottle to show how you can observe the movement of the air. He made us understand that ventilation may involve expertise, but it also involves common sense.

For another conference, I had Monona Rossol of Arts, Crafts, and Theatre Safety, who said the important thing to ask about ventilation is whether you are pulling the bad stuff away from your nose. When I see a gas station attendant pumping gas, at those stations where you are not allowed to pump yourself, I often ask if he stands out of the way of the smell, so that he doesn't breathe it in. "Often," they say, "Oh, I never thought of that." One time, the guy said "It doesn't matter—it's too late—I've been breathing it for years." But then after a moment, he stepped away (it is a matter of where you can stand, there's always a spot where you don't smell the gas) and said, "Thank you."

Some ways of reducing the level of insult are easy, like stepping away from the vapors or wearing a mask and gloves. But some green transition topics are quite complex. You have to make sure the safer materials also do a good job. You have to worry about durability and water and all the factors that make things work as intended. You have to think in terms of life cycle.

It is not just a simple matter of not having certain bad things from a Red List, but also having better things from a Green List, and understanding how they come together functionally.

At this point in our early development of a new Golden Age we often have inadequate information. This is an evolving area,[42] and even though it is growing fast, it needs to grow much faster. We want to build things right from the beginning so we can concretely embrace our principles—live within our

42. One example of a tool for the evolution of safer products is Clean Production Action's *Green Screen*, https://www.greenscreenchemicals.org/learn/full-greenscreen-method. Its primary creator, Mark Rossi, won the NPPR's P2 Ambassador Award in 2020, https://www.p2.org/page-18105.

best intentions. But when we build it right from the beginning, we also avoid the waste of having to rebuild again later.

TRANSPORTATION

Key to green communities is getting transportation right. It's also key to environmental justice and our energy goals. There is no shortage of resources available to explain how we can use electrical power for many vehicles, and compressed gas (which can be from biological sources eventually), or hydrogen, and there are even designs for reducing impacts from airplanes. But even clean vehicles are a problem if there are too many, making walking impossible, requiring roadways that cut through neighborhoods, or roaming all over natural areas, as Land Rover commercials heedlessly celebrate, appealing to people's love of natural beauty while ignoring that it won't last if we keep driving over it.

Historian, critic, and visionary Lewis Mumford described decades ago how cars had become the primary shaper of cities, destroying their essential purpose of bringing people together. He points out that we think we can satisfy the demands of the automobile by building more roads, but there is an iron law, that the roads will simply fill up, and the car will demand more. Mumford insisted on the need for regional planning and pointed to the Netherlands as an example. Their development of the railroad *supplemented, instead of supplanting*, the existing waterways, and then roadways were planned as supplements as well. This contrasts sharply with the US history in which railroads supplanted canals and then roadways supplanted railways. Mumford says, "If we are to save the hearts of our cities, the choice is clear and urgent. Does the city exist for people, or for motorcars? There is only one practical answer, we must boldly control the use of the motor car in the dense core of our cities. The heart of the city should be served chiefly by rapid transit, buses, taxis, and above all, the human foot."[43]

Yet regional planning has historically been frustrated by the authorities of towns and cities and the philosophy of leaving development to the market. But who, asks Tom Angotti, professor of Urban Policy and Planning at Hunter College, will do the planning? He notes that a regional authority can privilege suburbs over poorer populations in inner cities:

43. "Lewis Mumford on the City, Part 2, The City, Cars or People," a 1963 National Film Board of Canada production by Ian MacNeill: https://www.youtube.com/watch?v=ejeXiC4leiY.

> Suburban sprawl is wasteful and inefficient. It isolates communities and is bad for the environment. Most Americans live in large metropolitan regions and most planning is municipal, fragmented and parochial. True enough.
>
> But who will do the planning for the region? The majority of people in metropolitan regions live in suburbs. Will they remake the whole region in their image? Will powerful downtown business interests join with them to remove all obstacles to the global marketplace, including communities with poor people and people of color? Will regional planning put more power in the hands of those who have produced the disorderly sprawled behemoth in the first place?[44]

Robert Caro's portrait of Robert Moses is a story of regional development that was dominated by one man. He limited the ability of people of color living in poorer areas of New York City to access the beaches of Long Island "by building bridges across his parkways too low for buses to pass under."[45] Moses cut neighborhoods in half by building highways through them, and would have done the same to Greenwich Village, SoHo, and Little Italy, some of the most celebrated and cherished areas of New York, had he not been stopped by people like Jane Jacobs,[46] who worked against tactics such as short or obscure notice of public meetings, designation of a neighborhood as "blighted," and organizations masquerading as representative of residents. Jacobs later articulated a vision of urban life as a place for enriching cultural exchange and diverse uses in her 1961 *The Death and Life of Great American Cities*. The Regional Plan Association had been siding with Moses. Regional planning, though part of an answer, is not in and of itself enough of an answer to the question of how to marry cities and the need for movement.

Unless, of course, the planning process is equitable and takes into account the needs and wishes of the residents. "Neighborhood-led development" is a phrase used to describe a variety of approaches that have the intent of putting the decision-making power in the hands of those who will be most affected by the design of future cities. For example, the START program in Indianapolis:

> The START Program was developed as a part of the Twin Aire Great Places 2020 designation. Over the past several decades, neighborhoods like Twin Aire have been disempowered through a process of disinvestment, disruption, and suburban flight. These neighborhoods have a higher representation of minority and low-income residents, two groups that have historically been kept out of decision-making processes, whether intentionally or through ignorance. The Twin Aire Great Places 2020 program is a way to flip that history by investing

44. "Regional Planning and Reason," March 6, 1998, on the website of the Planners Network, "the organization of progressive planning," http://www.plannersnetwork.org/1998/03/regional-planning-and-reason/.

45. *The Power Broker* (1974), 318.

46. https://www.theguardian.com/cities/2016/apr/28/story-cities-32-new-york-jane-jacobs-robert-moses.

in these communities first through a process of neighborhood led planning and implementation of resident-selected goals and objectives.

Related to equity is the greening of the urban space, which necessarily limits the pollution and fragmentation due to roads. In 2010, the Natural Resources Defense Council worked with the Congress for New Urbanism and the US Green Building Council to develop LEED for Neighborhood Development, or LEED-ND. LEED-ND values reuse of vacant city lots, respects local biodiversity and wetlands, and seeks to discourage sprawl—the design that necessitates cars, favoring walkability. Cities have used the criteria "to develop codes, ordinances, and incentives of their own. Sarasota, Florida, gives fast-track permitting to developments pursuing certification. Syracuse, New York, used it to dictate the planning of an arts district. Boston requires that multi-building projects prove they are certifiable before giving approval."[47]

Transit-Oriented Development is the obvious idea that you ought to build near public transportation, but you have to plan for the changes in parking capacity that will result if things work out as intended. Where I live it was once a sensible practice to drive only a few miles to the "T" where I could park and ride into town. But then the garage filled up so that when I got to my car after a hard day's work, it would take me twenty minutes to get out, the place was so congested. Although I discovered that I could park in a nearby residential area and walk through a swamp to the garage, enjoying birds and cattails waving in the wind, they built a boardwalk through the swamp and slowly people discovered this option and the streets of the residential area started to fill up, and soon parking was limited. You have to plan for when things start to fill up. It now made more sense for me to drive all the way into the city even though the best I could get there was a two-hour meter.

The issue of green transportation brings me back to my roots. I began this book by thinking about how I began thinking about how to transform society when I took control of my own course of learning at Hampshire College, founded on the idea of self-directed study, in 1970.

It seemed to me, then, that we were missing the chance to harness the rising energy of hot air, or helium, and that a simple system of cables to the top of the nearby hill could convert to horizontal direction the vertical force of the rising of a hot air or helium-filled balloon,[48] and that once on top of the mountain, one could use gravity to roll down the other side. A system of such cables could enable a country with well-spaced small hills to have clean energy public transport. It would be slow, but scenic. Would the system be an eyesore, sullying the view of our local mountains? At that time, I envisioned

47. https://www.nrdc.org/stories/why-nrdc-helped-develop-leed-ratings-neighborhoods.

48. I also learned at that time of the massive losses of helium to space occurring as a result of drilling.

this, crews were busy at work quarrying a small mountain just down route 116 from the college. For years it was a grotesque site of a giant half-eaten rock, until it was completely flattened. I was aghast that anyone had the right to take away what was once a beautiful landscape feature.

For a year or two, I toyed with that idea as well as elevated pathways alongside roads, for skaters; bicycles hanging from cables, and chairlifts going down the strip between two-way roads, powered by wind and elastics that stored the wind's energy. I dreamed of slides from the higher floors of buildings that might take you several blocks before you hit the ground. I put these ideas into the head of an inventor I imagined, who promoted them in the late 1800s, because he anticipated the rise of poisonous industries and hoped to avert them, aiming for a degree I called "Creative Writing in the Public Policy of Science." My science advisors loved my work, but the English professors did not all love my writing. I had turned to fiction because I was no good at the necessary engineering to make my case and, as it turned out, wasn't so good at fiction either. I felt like a failure, but I did manage to graduate with a degree that perhaps no one else in the world has, and a habit of envisioning the world I want to see come about.[49]

If the streets were made of tarmac,
And everybody roller-skated,
I would be so happy,
I would be so elated.[50]

Now people have skateboards with little batteries. They have electric bikes. They have hang gliders. People talk about bringing back blimps—saying that they got a bad rap because of the *Hindenburg*, but that was filled with hydrogen because we didn't want to give the Nazis helium. But blimps are much slower than jets, and people keep wanting those to be faster. And because of the lack of planning, we don't have a place for those amazing little scooters. Biking is dangerous because of cars and trucks, and the bike paths are popular for walking, so you can't go too fast on them, and people want to put speed limits on them or ban the powered bikes. The bike paths are typically made from old trainways, meaning you can never put that form of public transportation back there.

The need for the time is a feast of imaginings. What could our city look like? With parks and greenways, with walking primary, with nonpolluting transportation choices? Many solicitations for design ideas would produce many gorgeous ideas. We'd have urban walls green with trees and ivy. The city would resemble a forest to the slightly blurry eye.

---

49. Creative Writing in the Public Policy of Science.
50. First stanza of a song I wrote for a class in law school.

No matter how hard it is to transition from our dirty practices it will be easier with many visions to choose from. If something is fun, people do it. Providing the incentive of reasonable profit, and honor, for the designs that best meet the needs of the future and the living environment will also foster evolution of the caring arts of responsibility. I dreamed of a world where people rolled quietly by and that made me happy, and now that they do, it gives me hope. I am grateful to those whose dreams were powerful enough to come to life.

Chapter Eleven

Endnote

Our recent history happens to be one in which government has been in bad repute, very largely because of propaganda campaigns conducted by immoral profiteers. Opportunists like Donald Trump have taken advantage of that to gain power, and have torn down instruments of policy that the taxpayers paid to erect, destroying value painstakingly created, as if that increased any aspect of freedom other than the freedom to be irresponsible and selfish. The basic foundation of environmental law has withstood the assault, but it is badly damaged. When we build it back, as President Biden is doing, we can, as he says, "build it back better." It is an aspect of freedom to use government to protect our environment and public health.

Throughout my career I have had the pleasure of telling people about the work I do, and how feasible it is to reduce the level of insult to our bodies that environmental problems are. I tell them stories about how companies found ways to be cleaner, safer, more responsible. I talk about how there are good companies out there and we, as the people, can use our government to favor them, and gently nudge the dirty, irresponsible companies from their positions. They gaze at this vision with wonder, having rarely considered it before. This book is written to continue that conversation.

So much is left out of this book: it is not intended as the canonical overview of relevant matters. There is, for example, the recent reauthorization of the Toxic Substances Control Act, to me a stark example of how the public needs to know more about these issues, as it took vast steps back as it took steps forward. That book should be written. This is not it.

Or more of the history of how we saw miraculous changes in industry, how companies would respond when you simply asked them questions—how we found alternatives to using toxic materials, wasting energy, water, materials.

Or more on cooperative alternatives, shared intellectual capital, what I call "new orgs" that could replace exploitative enterprise.

So many things not covered. For example, the proposals of the National Advisory Council on Environmental Technology,[1] the Aspen Institute, or the many other advisory groups and pilot programs at the state level. But history is not just told in surveys or overviews. It is also encapsulated in personal experience, even by those who may not be considered significant, if they have witnessed great events. I have achieved none of the usual signs of significance—I never rose above mid-level staff level, and have only ever been a lecturer, not a "real" professor, though students have called me that. I now have a full-time contract at Boston University. But I have published a number of works, hosted hundreds of public events, created innovative programs, won awards,[2] and convinced people to change who did not think they could. I have had the pleasure of seeing my ideas become realities and prove out that the ideals that motivated them were practical. This idea—of the potential practicality of ideals, and the terrible impracticality of abandoning them—is to me the most important thing I have learned. It confirms my general tendency to hope for the best. I believe that if you work in that hope, you make that hope work.

So many other environmentalists have had similar careers. There are many like me out there. I met them, worked with them, witnessed their good deeds. They were in agencies, in nonprofits, on their own as concerned citizens, and often, thank god, in companies, and sometimes in high positions. I wish I had met more, of course, but they *were there—they are there*—all over. We need to find and reward those who do good work. The next time you hear someone refer to those who work in government as "hacks," or someone who couldn't get a better job in the private sector, please give them a look that shows you regard them as overgeneralizing in a destructive and unjust manner. And the next time you are tempted to blame an agency's staff for failures try to find out if they are being controlled by political appointees, and raise your voice as a citizen in defense of the concept that the agency works for the people, and must not be captured by the politics of the party in power. During the Trump administration, I had to tell people countless times that if people within EPA spoke out against his policies they would simply be fired or removed from positions where they could continue to defend existing programs to some

1. Some have a similar message as this book, such as *Transforming Environmental Permitting and Compliance Polices to Promote Pollution Prevention*, EPA-100R-93-004, April 1993, Technology Innovation and Economics Committee of NACEPT, chaired by Nicholas Ashford, which described in scope and detail how government can support clean technology innovation to benefit both the economy and the environment.

2. Among them Al Gore's Hammer Award, EPA's Merit Award, the National Pollution Prevention Roundtable's P2 Champion Award. I was a planner of projects that won two Ford Foundation Awards for Innovation in Government.

extent if they kept quiet. This is hard for many to understand, but it's a simple concept: the agencies should not work for the executive, but for us, and the executive's job is to make sure they do that. To blame the entire agency and decide it's no longer useful is exactly the outcome the opponents of environmental progress seem to have wanted. We can resist that outcome by understanding the sources of our political disunity and making a real effort to understand what's going on.

There are many reasons to have hope, but the best is that we can each find how we can contribute, and join the worldwide movement to develop the societies that life itself demands of us. We can all belong to the beginning of that world.

Appendix: Addenda to the Agenda

This chapter is a sample exercise, suggested not just for courses on environmental issues or formal training of environmental teams, but for one to do on an individual basis. Imagine the Agenda—the perhaps unwritten list of things that people use to determine what to do next, how to invest time, money, energy, commitment, and thinking. That certainly includes climate change and, increasingly, environmental justice. It has always included clean air, water, and soil. What needs to be on the Agenda? We can all contribute to that discussion better if we spend time imagining that we individually have a chance to lend our voice.[1]

I found myself thinking about what ought to happen, constantly, as a form of therapy, ever since I first had the sense that things weren't quite right with the adults of the world. That sense came upon me as a young boy when I first learned about atomic bombs, watching a Jerry Lewis movie. How could such a thing be, I thought? I would see the Herblock cartoons in the newspaper, big nuclear bombs with evil grins. It seemed so clear to me that Nikita Khrushchev wouldn't want to kill me if he got to know me. That thought was consoling, and on it can be constructed proposals for greater interactions between peoples who might otherwise fall into conflict.

I have found that asking my students to come up with proposals for Addenda to the Agenda is something they very much enjoy doing.

---

1. And is a useful tool for countering what Roger Cobb and Mark Ross term *Cultural Strategies of Agenda Denial* in their 1997 study of "when opponents succeed in portraying initiatives as threats to widely held world views and identities," https://kansaspress.ku.edu/978-0-7006-0856-0.html.

ESTABLISH AN OVERALL GENERAL METHODOLOGY

It might help to think of a plan for progress that consists of four steps:

Step One: Constructive Engagement and Visions of Consensus.
Step Two: Options Analysis.
Step Three: Comparative Analysis and Collaborative Decision-making.
Step Four: Implementation, Evaluation, Reiteration

PRIORITIZATION

I have been in many meetings where people asked, "What are the most important priorities to address?" And commenced with discussion of what presents the most harm and/or the most risk of harm. This is critically important to do, but it is also important to consider what is most doable, especially when there is a general reluctance to grapple with environmental and public health issues, as there has been for decades now. In this situation, we need to show that we can succeed, and that facing our problems will pay off. It can make sense to start with something small, whatever gets us into the mode of world repair.

INVESTMENT, NOT COST

One can read a list of things to do and say, "Well, that's all well and good, but where does he think we are going to get the money?" This is just simply a progress-killer and can be applied to anything we do. Why privilege the status quo, as if everything is just fine?

Sometimes this question reveals an implicit assumption that environmental and public health expenditures are "extra," but they turn out to be basic when you look at them closely. The characteristic of basic means investment in maintenance is wise. Without healthy bodies and ecosystems, we have nothing.

Intelligent spending will have substantial paybacks. If you substitute cleaner practices for polluting, the benefits continue to accumulate over time. If you prevent a species from becoming extinct, you prevent a loss that is irrevocable. Instead of asking where we will get the money to pay for environmental initiatives, we should ask how we can conceivably afford not to invest in reducing our impacts. We've been asking the wrong question for a long time.

When you see the extraordinary value of environmental and public health investments the question shifts to how can we put off doing them any longer. We need not fear that the list of things to do is long, because each act of sustainability, by definition, builds the common wealth, as each item pays back. We need, instead, to fear that which is unsustainable, to see how it is like our lives running down the drain, and learn how to hold on to what counts.

We each have our list of what needs addressing: climate change is now at the top of many around the world, along with extinctions, desertification, wildfires, wastes (see the planetary gaps discussion earlier in this book). The list is long but we need to have it. We need to keep adding to the list the new concerns we see, no matter how short we would love that list to be. The idea of the Addenda to the Agenda is one of a continuous, ongoing effort. Here are just some things that I propose for the list.

BETTER IMPLEMENT THE POLLUTER PAYS PRINCIPLE

A stunning example of how our system of environmental governance has been captured and warped by corporate interests is the elimination of the Superfund Tax during the time of the Gingrich Congress.[2] The original concept of the Superfund was that the oil and chemical industries are the source of toxic chemicals that end up contaminating land, so those who made money from selling them should contribute to the necessary cleanup. The problem is that because the fund this tax supported is now too small to pay for what we need, EPA must aggressively go after potentially responsible parties to get the money for cleanup. Companies that have sent their wastes to a site that mismanaged them now have had to pay not just for their own share of the mess but also extra to make up for those who can't be found or are out of business or money. This is termed "the orphan share." A well-funded Superfund would enable EPA to pick up that share, and this would lessen the understandable resentment businesses feel toward the government that is just trying to clean things up. In 2014, Senator Cory Booker and other New Jersey legislators proposed reinstating the tax:

> You make the mess, you clean it up. It's as simple as that," said Sen. Menendez. "If you profited from polluting the environment then you should pay to clean up the mess you leave behind. It's time to restore the Superfund Trust Fund, make it solvent, make it effective, and hold polluting companies accountable

---

2. After the Deepwater Horizon oil spill EPA called for the tax to be reinstated. "EPA Backs Polluter Tax to Pay for Superfund," WebCPA, June 22, 2010, https://www.accountingtoday.com/news/epa-backs-polluter-tax-to-pay-for-superfund.

for the damage they do to our environment. It's time to clean-up New Jersey's 113 Superfund sites." . . .

"The federal Superfund is running on fumes, which has real implications for the people of New Jersey. It is important to reinstate the lapsed Superfund fees as soon as possible," said Judith A. Enck, Regional Administrator, US Environmental Protection Agency.

The Superfund Polluter Pays Restoration Act of 2014 reinstates the excise tax on polluting industries to pay for the cleanup of Superfund sites, relieving taxpayers of the expense. It also expands the definition of crude oil in order to make oil from tar sands and shale subject to the excise tax. Additionally, it makes funds available to the Environmental Protection Agency (EPA) on an ongoing basis, not subject to annual appropriations.[3]

The tax has now been reinstated by the Biden administration.[4] We may now hope it can be used to pick up the orphan share and thus reduce resistance to cleanups. We may now hope it will help convince chemical companies to work harder to find cleaner alternatives to the products that have been found contaminating our land.

CREATE CORPS OF TREE-SAVIORS

We are familiar with tree-planting efforts. Arbor Day is dedicated to that, FDR promoted it,[5] China has launched a tree-planting effort, Wangaari Maathi initiated the Green Belt across Africa.[6] But it must be done intelligently. Switching to shrubs and plants locals can use might work in desert areas better than trees that suck up large amounts of water. Some plantations reduce biodiversity and produce a terrible sameness, while others provide habitat for a variety of species, allow for agriculture in the shade, and enrich the spirit. Besides replanting trees (reforestation), planting them where they have not grown for a long time (afforestation), there is agroforestry, which

---

3. December 14, 2015, press release: https://www.booker.senate.gov/news/press/2015/12/14/booker-menendez-pascrell-pallone-announce-legislation-to-hold-polluters-responsible-for-cleanup-of-superfund-sites.

4. https://www.washingtonpost.com/climate-environment/2021/12/17/superfund-cleanup-biden-infrastructure-bill/.

5. See Douglas Brinkley's 2016 *Rightful Heritage: Franklin D. Roosevelt and the Land of America*, detailing how extensive and important were the second Roosevelt president's conservation efforts and his special interest in trees, reflected both on his own property and by his New York State governorship, as well as in the reforestation of this nation after decades of development had denuded much of it.

6. This resulted in the planting of tens of millions of trees, https://wangarimaathai.org/wangaris-story/.

involves growing crops with trees, such as shade-tolerant plants within the canopy or using trees as windbreaks between fields, and there is urban forestry. With the increase in storm intensity, assaults from invasive species, changes in the atmosphere and water supply, and growth of vines, trees can be weakened and fall. We need to train and hire tree caretakers to assess their health, treat them when invaded by pests, feed and water them as needed, perhaps stabilize them, and cut the vines that are choking them. When they are blown down, some can be lifted back up and re-rooted, but we need a rapid response and knowledgeable tree-saviors to find them right away, assess if there are enough roots left to survive, and who know how to "right" the tree. "Together with other land-management practices, afforestation and reforestation could remove about one-third of the carbon needed to reach the 2030 Paris Agreement climate goals."[7]

USE SPACE TO SAVE THE EARTH, NOT TO LEAVE IT

Earth observation from space is an increasingly important tool for understanding what is going on down here, from measuring climate to understanding the extent of deforestation and ocean algae blooms to animal migrations. The NASA Earth Observing Systems have applications in Agricultural Efficiency, Air Quality, Carbon Management, Coastal Management, Disaster Management, Ecological Forecasting, Energy Management, Homeland Security, Invasive Species, Public Health, and Water Management.[8] Just think about how helpful that is compared to going to the moon or Mars. What does that accomplish? Maybe it will prove to be of great value, but it's speculative. Earth observation has immediate, substantial, indeed *urgent* importance. Space launches have serious environmental impacts, particularly on the ozone layer that we have worked so hard to keep intact.[9] We can't afford to have space tourism. We can't afford to go into space to mine asteroids and or the moon, when we haven't yet learned to mine our own waste streams. Let's just think twice before we get all starry-eyed. Earth Observation is an extremely important use of our ability to go into space. If we clutter up the sky with frivolous or commercial purpose, we might not have room for what we really need.

---

7. *The Economist*, Intelligence Unit, https://carbonremoval.economist.com/afforestation-and-reforestation.
8. See NASA's Earth Observing System Project Science Office, https://eospso.gsfc.nasa.gov/.
9. J. A. Dallas et al., "The Environmental Impact of Emissions From Space Launches: A Comprehensive Review," *Journal of Cleaner Production* 255 (May 10, 2020): 120209, https://doi.org/10.1016/j.jclepro.2020.120209.

ASSIST COMMUNITIES

Small or low-income communities have trouble providing clean drinking water, sanitation, recycling and ordinary trash management, household hazardous waste management, electronics waste management, and so on. They could be providing their residents with bulk energy purchasing to get them the best price, but only if they have the staff to figure it out. They could be planning microgrids with solar panels and battery storage so that critical services and residents too would have reliable, resilient electrical systems. They could have health and fire departments that keep people safe—that actually inspect places with dangerous chemicals to make sure they are stored and managed safely—and they could have quality schools, but often small communities have trouble supplying these things, even when they are mandated. We expect communities to pay for much of what one would consider basic needs through local taxes, but when people are not wealthy and there are few businesses small and low-income communities cannot meet these needs. This is where county and state governments and the federal government ought to step in so that all citizens of this country enjoy protection of fundamental rights. The system in which we expect poor communities to pay for all the things that rich communities can afford is not a system that affords that kind of protection, but our founders formed a strong central government for the purpose of creating one country. The Constitution, ratified in special conventions allowing more people to vote than ever before, ordained and established a new government for the purpose of establishing justice, insuring domestic tranquility, providing for the common defense, promoting general welfare, and securing the blessings of liberty for citizens and their posterity.[10] None of these things are accomplished with a system that leaves poor communities to fend for themselves.

REALIGN DEFENSE WITH DEFENSE

We now use a terrifying offensive threat as if it is defense, but it is not. It is a blatant attempt to frighten the rest of the world into submission. This does not work. History shows this amply. The Vietnam War was an attempt to intimidate the communists from entering South Vietnam through "Rolling Thunder," an intense bombing campaign. It only increased determination. Had we used the nuclear weapon it would have simply turned the world even more fiercely against us.

10. The Preamble to the United States Constitution.

The threat of such complete destruction damages the user as much as the recipient, even when not used. We have wasted enormous amounts of money and the chance to win over the world and each day we fail to work on nuclear disarmament is a loss to humanity.

The wars of choice in Iraq and Afghanistan have not worked out well. Why do we threaten the world with the prospect of force, maintaining hundreds of bases around the world?[11] A big question. Surely one aspect is the justification of protecting shipping, international trade. But could we instead lead the development of an international shipping protection force to which all nations contribute? If so, then it seems we ought to be able to rededicate our military to actual defense.

A key moment in the shaping of our current situation was when the US decided to develop the hydrogen bomb. According to an account by atomic physicist Hans Bethe that was not declassified until 1980, the General Advisory Committee (GAC) conducting its meetings in secret concluded: "We all hope that by one means or another, the development of these weapons can be avoided. We are all reluctant to see the United States take the initiative in precipitating this development. We are all agreed that it would be wrong at the present moment to commit ourselves to all-out efforts toward its development." Bethe comments on the potential of a proposal by some members of the committee:

> The proposal was to enter negotiations with Russia with the goal that both countries commit themselves not to develop the H-bomb. If such an agreement could have been reached and had been kept, it would have gone far to avoid the peril in which the world now stands. At that time neither we nor presumably the Russians knew whether an H-bomb could be made. In this blissful state of ignorance we might have remained for a long time to come. Since the technical program was a very difficult one, it could never be accomplished without a major effort. It is possible, perhaps likely, that the Russians would have refused to enter an agreement on this matter. If they had done so, this refusal would have been a great propaganda asset for us in the international field. . . . It is difficult to tell whether or not the Russians would have developed the H-bomb if we had not. I am not sure what would have happened if we had followed the recommendations of the GAC majority and had merely announced that for such and such reasons, we would refrain from developing the H-bomb. Once we announced that we would go ahead, the Russians clearly had no choice but to do the same. In the field of atomic weapons, we have called the tune since the end of the war, both in quality and in quantity. Russia has to follow the tune or be a second-class power.[12]

---

11. See, for example, the books of Chalmers Johnson, *Blowback*, *Nemesis*, and *The Sorrows of Empire*.

12. "Observations on the Development of the H-Bomb," Appendix II in the 1989 version of *The Advisors, Oppenheimer, Teller, and the Superbomb*, by Herbert F York (1976), https://fas.org/nuke/guide/usa/nuclear/bethe-54.htm. The public did not know at the time that the GAC majority had recommended not going ahead with the bomb. The deliberations were secret.

The United States went ahead and developed the insanely terrifying and completely useless hydrogen bomb. The Marshall Islanders lost homes[13] and we have frightened the entire world to death. The prospect of such devastating destruction of civilization has only produced fear, resentment, and imitation. We can take the enormous defense budget and repurpose all of that civic dedication that our military personnel display, and have something we can all be proud of. The world will appreciate us more and we will be safer as a result.

INCREASE FOREIGN ENVIRONMENTAL AID

Truly helping other countries will help us. While corporations may be in competition with other countries, we should not be. We should be seeking to build relationships of mutual support. Even nativists who don't want an influx of refugees should be for helping people in other countries, so they can afford to stay where they are. For those who worry about disease from other countries, helping the world fight disease will reduce its incidence here. For those who worry about other countries invading us to take our resources, helping them to use their own wisely will reduce that threat. Helping other countries promotes friendly exchange and creates ties that help keep the peace during times when particular interests try to force issues.

The world needs help in fighting climate change, reducing deforestation, protecting ecosystems, reviving wildlife, protecting soils, generating food and livelihoods. We should regard ourselves as part of the world and in a position to help others as members of a group project. This could be an obvious cause for pride in this country, which can give a boost of energy to an effort. For those who already believe that we are the most generous country in history, we can't rest on the laurels of past acts and we also have mistakes to make up for. The world may no longer want us to be its leader, partly because of those mistakes. For many good reasons, it will be good for us, too, not to think the world is ours to run but to learn to just be part of it. As we step down from the terrible position that we thought we had to preserve, we can do so in a way that leaves a sensible world organization in place, of commerce, relationship, and law. We are still in a position to influence the creation of a new engagement at all levels. Our foreign policy has been devoted to specialized industrial, political and other interests but can be reoriented to reflect the heart of its people, which should have no need to prevail over other economies, but could seek to cultivate relationships.

---

13. https://mh.usembassy.gov/the-legacy-of-u-s-nuclear-testing-and-radiation-exposure-in-the-marshall-islands/, https://www.sfgate.com/opinion/openforum/article/Remembering-the-Marshall-Islands-2516796.php.

REDUCE ELECTROJITTERS

As discussed earlier (chapter 6), while some alarm about the effect of electromagnetic fields (EMF) on humans is surely overblown and represents fear-based conspiracy theory, some is backed up by science. Nearly all commentary that is not biased notes the need for more research. For example, a recent briefing to the European Parliament found:

> *Summary* The fifth generation of telecommunications technologies, 5G, is fundamental to achieving a European gigabit society by 2025. The aim to cover all urban areas, railways and major roads with uninterrupted fifth generation wireless communication can only be achieved by creating a very dense network of antennas and transmitters. In other words, the number of higher frequency base stations and other devices will increase significantly. This raises the question as to whether there is a negative impact on human health and environment from higher frequencies and billions of additional connections, which, according to research, will mean constant exposure for the whole population, including children. Whereas researchers generally consider such radio waves not to constitute a threat to the population, research to date has not addressed the constant exposure that 5G would introduce. Accordingly, a section of the scientific community considers that more research on the potential negative biological effects of electromagnetic fields (EMF) and 5G is needed, notably on the incidence of some serious human diseases. A further consideration is the need to bring together researchers from different disciplines, in particular medicine and physics or engineering, to conduct further research into the effects of 5G. The EU's current provisions on exposure to wireless signals, the Council Recommendation on the limitation of exposure of the general public to electromagnetic fields (0 Hz to 300 GHz), is now 20 years old, and thus does not take the specific technical characteristics of 5G into account.[14]

In the United States, the Federal Telecommunications Act recognizes a responsibility to provide access to the modern world of communications to all, but recent regulations by the Federal Communications Commission (FCC) designed to speed up installation of wireless equipment has created opposition on the local level, as it limits their ability to control the placement of a significant amount of infrastructure. The Environmental Health Trust, a science-policy group headed by Devra Davis, sued the FCC in 2020 to base its rulings on recent science, instead of the twenty-four-year-old science it used in developing its wireless rules.[15] Some claim that fiber-optic can do much that 5G is intended to do, and is in some respects better (including

---

14. *Effects of 5G Wireless Communication on Human Health*, https://www.europarl.europa.eu/RegData/etudes/BRIE/2020/646172/EPRS_BRI(2020)646172_EN.pdf.
15. https://ehtrust.org/eht-takes-the-fcc-to-court/.

without the hazard of electromagnetic fields), and it seems an integrated system utilizing both will best and most safely provide what people want.[16] The environmental system of the future would hopefully employ a less draconian approach than that approved by the FCC, which removes local authority and opens the field to speculative development of towers feared and hated by many residents. Because a town can impose some controls, if it has adopted an ordinance before a developer proposes installations, the current rule favors those communities with adequate staff and governance to develop such an ordinance, and disfavors those that do not have the capacity to do so, or to even be aware of the need until it is too late, for the rules require a response within a short time frame.[17]

An effective system of environmental governance would rapidly increase our understanding of potential impacts before allowing such an uncontrolled build-out. The business publication *Forbes* published a post in 2019 by risk specialist Jayshree Pandya, which notes:

> Now, in a rapidly transforming digital age, with wireless routers, Wi-Fi and Bluetooth, 5G, the internet of things, radio and TV signals, mobile phones, and more, the applications and sources of EMF are proliferating. As a result, fears are amplifying about possible connections between EMF and adverse health effects on humans, other living beings, and the environment. Based on the thousands of published reports, there are growing concerns about radiofrequency radiation that can (i) damage the DNA of any biological species and lead to cancer; (ii) cause oxidative damage that can cause premature aging; (ii) disrupt cell metabolism; and (iv) potentially lead to other diseases through the generation of stress proteins.[18]

Dr. Pandya asks: "Since biological effects are measurable responses to an electromagnetic stimulus or a change in the environment, should we not put efforts towards having a sufficient understanding of the impact of human-made electromagnetic fields on biology on a cellular level?" Boosters may ridicule those who oppose modern technology as manufacturers of fear and believers in junk science. But proponents of the rapid expansion of wireless

---

16. https://scientists4wiredtech.com/2018/10/fairfax-to-study-fiber-optic-broadband-instead-of-5g/. This *Forbes* article finds that "What is crystal clear (pardon the pun) is that fiber optic networks are the path forward to getting terabytes of data to consumers' homes and businesses instantly." https://www.forbes.com/sites/theyec/2020/08/05/the-role-of-fiber-optic-networks-for-the-future-of-5g/?sh=142caa4fb411; for a technical comparison of the two, concluding that they should coexist in an integrated system, see https://www.rfwireless-world.com/Terminology/5G-vs-Fiber.html.

17. For example, a town can ensure the devices are measured to meet established limits, even though those limits are considered too low, can limit how trees are cut, can ensure they don't make noise, but not within the very short time provided if it does not already have an ordinance in place when a developer makes application.

18. https://www.forbes.com/sites/cognitiveworld/2019/09/09/an-electromagnetic-health-crisis/?sh=4b63029630e5.

technologies are relying on outdated science that didn't even encompass examination of the fields to which we are being exposed today, or will tomorrow. They are relying on the absence of science to promote an industry that should not be developed through imposition. We don't need to look back on this moment with regret that we could have built safer systems to provide access to modern communications. It will be far more costly to rebuild than to get it right at the outset. Carpenters know to measure twice in order to cut once. *We are not even measuring once*. Electronic communications systems are technically complex but there is a simple principle of care here that anyone can grasp, and which is being ignored. The 5G build-out seems to be another example of technological momentum and profit-seeking mania, and we would better off to calmly take stock of risks, rights, and safer alternatives in a reasonable process.

PRESERVE THE ENVIRONMENT FOR FUTURE PEOPLE

Our existing conservation laws, such as the Multiple Use Sustainable Yield Act and the Federal Land Policy and Management Act, require the federal government to balance competing uses, such as recreation and resource use, with ecosystem protection. Similarly, the National Environmental Policy Act that requires assessment of potential impacts fails to fully achieve its purposes, in part because it does not take sufficient notice of cumulative impacts, nor has it been interpreted to require prevention of environmental harm—just the assessment of what it will be. (Much greater weight on the consideration of alternatives would help.) Environmental law experts have proposed an Environmental Legacy Act, which would create a mechanism to ensure we leave a healthy environment for future generations to enjoy. The act would limit resource use to that which can be renewed. The current system's vesting of wide discretion in politically influenced agencies can be described as allowing a "spend-down" ethic that privileges the needs of today over those of future generations. The act would mandate that responsible agencies monitor and allow no degradation or depletion over permissible limits and in particular to evaluate ecosystem resilience. A citizen-suit provision would help ensure the agencies do their job.[19]

A seminal article articulating the concept of intergenerational equity is the 1992 article "In Fairness to Future Generations and Sustainable Development" by legal scholar Edith Brown Weiss:

19. The Environmental Legacy Act is described in Alyson Flournoy and David Driesen's 2010 *Beyond Environmental Law: Policy Proposals for a Better Environmental Future.*

> Sustainable development relies on a commitment to equity with future generations. This ethical and philosophical commitment acts as a constraint on a natural inclination to take advantage of our temporary control over the earth's resources, and to use them only for our own benefit without careful regard for what we leave to our children and their descendants.[20]

BUILD COMMUNITY SOLAR ON CONTAMINATED LAND FOR NEIGHBORS

While solar panels are dropping in price and becoming more and more affordable every year, low-income residents still have trouble affording them. In addition, anyone who is renting has to receive the permission of the landlord to place solar panels on the roof, and then the reduction in electricity costs may not be equitably shared, as the tenants may not have any rights in the matter. Those who cynically oppose renewable energy have used this situation to characterize the solar movement as one in which only the rich benefit, and because they can reduce their rates the cost of providing electricity increases for everyone else, impacting especially on the poor, unless constraints on raising rates come into play. As utilities have acted to slow the development of solar that eats into their current methods of making money, this disparity has worked for them. Community solar is the ability to participate in a project that is not on your roof, but somewhere in your community, however defined. It is a way for anyone to participate in the solar revolution ("Solar for All") and enjoy lower electricity costs as well as the knowledge that they are contributing to the reduction in greenhouse gas emissions. Much of community solar has a large off-taker which is not local residents. But it can be designed and encouraged with incentives to primarily benefit the neighborhood.

A related issue is that large solar projects are sometimes built at the expense of green space, or by cutting down trees. If they are to benefit lower-income communities in urban areas, those spaces and those trees are even more precious than usual. Many areas are blighted with contaminated land from former industrial operations. These are in inner cities and in the mining regions of the West. Where the contaminated land can host solar, they can provide energy for "pump and treat" operations that keep plumes of contamination from spreading in groundwater, for monitoring of the sites, and for supplying electricity for free to neighbors to compensate them for living near the contaminated site.

---

20. *American University International Law Review* 8, no. 1 (1992), https://digitalcommons.wcl.american.edu/cgi/viewcontent.cgi?article=1498&context=auilr.

Students in my research class found numerous sites large enough for solar farms in my state and were welcomed to come to the State House and report their findings to the staff of the relevant joint legislative committee. State and municipal help are provided, but more is needed for the market to be able to fulfill this opportunity.[21] More innovative solar placement, and concerted efforts to cover roofs, parking lots, and sidewalks (to reduce urban heat) instead of solar farms will help.

USE THE LITIGATION TRUST

Jan Schlichtmann is famous as the attorney in *A Civil Action*,[22] who ended up in bankruptcy litigating a toxic tort case on behalf of Woburn residents who drank contaminated water (*Anne Anderson v. WR Grace*).[23] After a period of years, he returned to Massachusetts and resumed the practice of law, and after learning about how he was going about it, I asked him to speak at an event I organized for people interested in innovative environmental legal developments. He talked about the litigation trust concept. Describing his approach in a suit against the Massachusetts Public Transit Authority, *Boston Magazine*'s Paul Kix wrote:

> In the two decades since Anderson, Schlichtmann developed and then refined—some might say perfected—a technique that keeps his cases largely out of the legal system: He establishes a public trust through which the assumed plaintiffs air their grievances and the facts they've uncovered (where Schlichtmann comes in) directly with the assumed defendants. The idea is to mediate before the sides litigate, and when it's done well, Schlichtmann turns to the court only to approve the settlement.[24]

Schlichtmann expressed several other purposes of the trust when he talked to my audience. He said that in his Woburn case, he was supposedly representing every plaintiff individually, but in fact he was really suing on behalf of the group, and it was a "legal fiction" that he could represent them individually. By forming a trust to which they all belonged, he could represent the trust and not have to engage in any fictional concept. In addition, forming the trust enabled the group to determine its goals in a fair and reasonable process, so that he could fairly and reasonably do his job on their behalf. It also provided

21. www.bu.edu/rccp in Spring and Fall 2018 projects.
22. The 1995 book by Jonathan Harr was made into a movie in 1998, starring John Travolta as Schlichtmann.
23. https://serc.carleton.edu/woburn/Case_summary.html.
24. Paul Kix, "In the Shadow of Woburn," *Boston Magazine*, September 22, 2009, https://www.bostonmagazine.com/2009/09/22/in-the-shadow-of-woburn/.

for a reasonable process of receiving and distributing the funds if the effort succeeded. He said he had seen plaintiffs take compensation and blow it in a short time, but the group could decide to take any recovery and invest it, and distribute it in a reasonable way that would not allow the immediate dissipation of funds. Here is a concept that helps fill the gap between individual suits and those involving many parties. If the legislation making it easier to sue those who recklessly place lead-containing articles in commerce when they know they will harm people discussed above becomes law, such a trust would be a way to go forward with such cases. All plaintiffs could agree ahead of time on a way that some of the recovery they seek could be used to prevent further harm, and not just compensation.

DO TORT REFORM TO BENEFIT PLAINTIFFS

In recent decades, the phrase "tort reform" has been dominated by efforts to benefit defendants, aided by stories about how plaintiffs are deluging the courts with frivolous suits. Tort reform has gone far to restrict what plaintiffs can receive as well as constrained their opportunities to even get into court or maintain standing to sue.[25]

We can use the concept of tort reform to mean rationalizing the law in terms of justice. Plaintiffs should have a decent chance to get deserved compensation because that will cause the evolution of responsible business. Malfeasance is curtailed when there is accountability. The bill discussed above concerning those who recklessly place lead in commerce is one example: tort reform should be about courts using market-share or risk contribution liability to apportion the costs of compensation among those who acted to cause the harm, even though the plaintiff can't trace causality of their specific harm to a specific defendant, because in the weighing of justice as between that group and harmed individuals, that is most just.

Another example is reforming the class action. Attorneys General could hire private law firms to take these cases and manage them to prevent attorney windfalls and ensure fair distribution and public benefits.[26] Similarly, the use of punitive damages strikes people as unfair, because a plaintiff can receive an outsize recovery far exceeding any real damages to them. Juries may be reluctant to use them for this reason, but they are an important tool for deterring bad behavior. Tort reform could include the use of punitive charges

25. See Sarah Staszak's *No Day in Court: Access to Justice and the Politics of Judicial Retrenchment* (2015).

26. See Judge Jed Rakoff, "The Cure for Corporate Wrongdoing," *New York Review of Books*, November 19, 2015, https://www.nybooks.com/articles/2015/11/19/cure-corporate-wrongdoing-class-actions/.

to benefit groups of victims similar to the plaintiffs, or the public at large, instead of the plaintiff who would be made whole by ordinary compensation. Courts could use special magistrates to direct their expenditure to benefit many. If we don't allow cases to be heard, we can expect harm to continue. Actions to punish plaintiffs for bringing frivolous claims are the direct method of countering that potential problem, not keeping them from being able to make their case, or limiting what they can recover to some arbitrary figure that may not be just.

Best of all would be to replace the tort system, when possible, for the kinds of injuries affecting many, funding it by charges on the harmful activity, so that it is easy to get compensation for harm, and the harmful activity is made more costly.

IMPLEMENT THE PUBLIC TRUST DOCTRINE

In 2014, the University of Oregon Law School's Mary Cristina Wood published *Nature's Trust*,[27] elaborating on Joseph Sax's concept of the public trust to argue that the government has a responsibility to protect our atmosphere.[28] In 2015, Julia Olson, representing children whose future is at stake if the atmosphere is not protected, filed suit in the case known as *Juliana v. USA*.[29] With the historic ruling by Oregon District Court Judge Ann Aiken in 2016, the concept was established at law that "the right to a climate system capable of sustaining human life is fundamental to a free and ordered society."[30] While the order has been reversed by the Court of Appeals it was not a denial of that principle. We should support such suits and continuously advocate for our governments to assert and enforce the public trust by protecting public assets, such as the atmosphere,[31] the water we drink, and the soil we need to grow our food. These are commonly owned and no other actor can fulfill the needed role of protector. It is up to us to insist that the government perform its

27. Subtitle: *Environmental Law for a New Ecological Age.*

28. "Three types of restrictions on governmental authority are often thought to be imposed by the public trust: first, the property subject to the trust must not only be used for a public purpose, but it must be held available for use by the general public; second, the property may not be sold, even for a fair cash equivalent; and third, the property must be maintained for particular types of uses." "The Public Trust Doctrine in Natural Resource Law: Effective Judicial Intervention," *Michigan Law Review* 68, no. 3 (1970):, 477, https://repository.law.umich.edu/cgi/viewcontent.cgi?article=4782&context=mlr.

29. https://www.ourchildrenstrust.org/juliana-v-us.

30. Opinion and Order in Case No. 6:15-cv-01517-TC, United States District Court for the District of Oregon, Eugene Division, November 10, 2016, p. 32, https://static1.squarespace.com/static/571d109b04426270152febe0/t/5824e85e6a49638292ddd1c9/1478813795912/Order+MTD.Aiken.pdf.

31. See Columbia University's Sabin Center: https://climate.law.columbia.edu/content/climate-change-litigation.

role as trustee owner of common assets and ensure they are not degraded or destroyed. Courts can tell this to the legislative and executive branches, and those branches should be instructed by voters to listen to the courts.

USE THE PUBLIC OPTION TO CORRECT MARKET SHORTCOMINGS

When we continuously and expansively fund the development of new, clean technologies that can compete with existing dirty technologies, we can take ownership of some of the intellectual property rights, licensing out the technologies to private enterprise and using the license revenue for continued investment in public interest enterprise. We should especially favor organizations that share profits and operate democratically. This approach, of public action in the commercial sphere, is consistent with the idea of free enterprise and competition, with the added concept that the public can be an actor bringing excellence and mutual benefit where the unregulated market does not. It is also an important concept to set against the idea of central control—the part of socialism that makes people afraid of using the tools of governance to influence commerce. Think of it as public free enterprise, without capture and unlimited profit-taking (for example, creating the internet and allowing companies to use it to dominate commerce, or developing fracking). It can be done so that the public retains the value its taxes paid to create, while the incentives for private enterprise remain reasonable.

When I was providing free advice to companies on how to comply with the law and how to do toxics use reduction, I was occasionally challenged by private consultants who asked what right we had to compete with them. I usually responded with several points, among them that we were not competing, because no one was providing the same service we were. Private consultants were typically hired to perform tasks such as filing permits and regulatory reports. We did not do that, but only told companies what they needed to do. I would note that this often generated business for private consultants. But another response was to ask questions about whether the consultant was helping the company to eliminate the cause of the regulatory coverage—if, for example, when they had to get a permit for air emissions of a volatile organic chemical, were they helping the company find alternatives to the chemical. This usually elicited the response that they tried, but failed, or that they didn't even try, which was not usually stated but obvious from their answer. A consultant who eliminates the problem at the source also eliminates return business. The service that we provided from the state, for free, demonstrated how toxics use reduction was possible and increased the demand from companies for that service from private consultants.[32] It was an example of how a public function can improve how private entities delivered their services.

LISTEN TO THE LIST

Once you ask yourself the question, what should happen, you might trigger something that keeps pouring out the "should." Where do you stop?

> We should enter and uphold all international environmental treaties, reversing our recent history of deferring to business interests. We should take the lead in forging treaties to stay within planetary boundaries and to establish economic incentives for cleaner world trade and the development of strong local economies providing full and fair employment worldwide.

Personally, I find it helpful to let them keep coming. Doing so increases the chance that I will have a response when there are questions about what we can do, what direction we should be looking in.

> We should expect our courts to defer to agencies entrusted with doing a job for us all, except when they are not; and to defer to the legislatures, except when they are not doing their jobs and fundamental rights are not being protected. Then we should not accept their deference to do-nothing, captured legislatures and administrations, but recognize their role as back-stop, guardrails, last-resort defenders of the basic components of a legitimate government.

I recommend it to others as a kind of helpful mental exercise because it makes you realize that if you could think of a good idea, then others can, too, and the chances that it will be someone who can put it into effect are not nil.

> We should work with other countries to preserve as much of the wild world as possible, so that it can regenerate.
>
> We should let wildlife and natural resource guardians represent the forms of life that cannot speak for themselves, and provide these protectors with funding.
>
> We should do everything we can to preserve all remaining indigenous cultures, as they are unique expressions of life, the rights of indigenous peoples merit respect, and each has something to teach the rest of us.

---

32. Some consultants such as Robert Pojasek and Robert Capaccio provided focused prevention work, but there were not many in Massachusetts until the Toxics Use Reduction program led to the training of TUR Planners.

But maybe no one is putting it into effect, so you have to say something. Now you have to think about how you are heard. Simply stating "shoulds" may not work. They may even cause resentment, backlash, hostility, and that, as we have seen, can be picked up by political opportunists and whipped into an ugly partisan attack on you.

So should you keep your "shoulds" to yourself? No one wants to be lectured to. "Shoulds" are annoying if you feel they are commands. Think about how you express them. If they are seen as wishes, desires, suggestions that others may share, if they are made in such a way as not to appear to be threatening to someone else's identity or beliefs, that can help. If they can be shown to align with people's identities and beliefs, all to the good. If they will interfere with plans that are at odds with the public interest, that doesn't mean we should avoid the topic, or saying what ought to be. It just means you have to give some thought and not simply express your insight about what would be right and good without consideration of how it might be heard, including how it might be misheard.

> We should fully fund water and wastewater treatment everywhere, for to fail to do this basic thing is shameful. We should stop discharging nutrients to the ocean and recover them for the soil.
>
> We should fund independent science and regard all industry-funded science as potentially biased.
>
> We should reorganize the regulatory system so that it addresses all media (air, water, land) at once, is cross-disciplinary, efficient, and tailored to the contexts of the regulated entities.
>
> We should make our system preventive, use all the tools of governance in a synergistic mix of strategies, and lead with assistance, using enforcement as a backup.

I have the fantasy that I am asked what we should do, and having this fantasy makes me think about what is the one key area to focus on. It is a puzzle I find far more enjoyable than the crossword which I used to do for the fun of the mental exercise. At the end of that mental exercise I recycled the newspaper. At the end of this one I have something I might be able to use. After some thought I answer my fantasied questioner: *Implement Relational Environmental Governance*. This includes:

> Ensure community engagement.
>
> Get the broadest input into rule development.

Use positive incentives to foster the kind of growth we need.

Elaborate the expectations of responsibility that apply to individuals and groups.

Continuously assess how well rules are understood and how they are actually working.

Continuously improve rule-making and implementation.

Simplify rules to ensure they are understood.

Establish partnerships to develop innovations that can have wide benefit.

Create public-private entities that provide public benefit.

Use preferences, subsidies, tax incentives to generate desired and needed activities.

Use the public option to set standards for the market to follow.

And when they ask, what else? I answer them (an imagined group): *We should reinvigorate the common law so that it serves as a deterrent to public threats and respects the rights of individuals. To do this, we should*:

Improve access to courts.

Create pathways for public interest advocates to obtain just settlements and distribute awards.

Create compensation systems where fault is too diffuse for fair allocation of liability.

Create fair allocation of liability where identification of specific cause cannot be made.

Eliminate the need for chain of causation proof for inherently dangerous substances and acts.

Provide adequate legal assistance for all.

We should revisit our basic environmental statutes and make them stronger:

Invigorate TSCA so that it discourages the use of toxic substances.

Invigorate the Federal Insecticide, Fungicide and Rodenticide Act so that it actually discourages the use of pesticides and herbicides.

> Invigorate the Food and Drug Administration and the Consumer Safety Protection Act so that they prevent incorporation of toxics into products we use and so that our Right To Know about what's in things is fully respected.

> Transform the Federal Communication Commission's Green Guides, which are supposed to discourage fake green advertising, into enforceable rules.

There's so many other items, and other ways to list what needs doing. How can this be packaged? Too long of a list stops being helpful. I ask students to think about how they can communicate a message of responsibility so that it seems attractive.

A PIE

To make it more attractive, we can call it the Project on Improving Environmental Governance (the PIE Project), whose motto would be:

> *Since they say we can't have our cake and eat it too, we're making pie!*

The PIE Project can have two parts: one, focused on improving existing regulations so that unnecessary burdens are removed and rules are made more effective, and two, evolving the new law that's needed to protect our biosphere.

The crust that holds everything together can be the concept of rights as the universally shared constant, like the speed of light, the same in all contexts, all frames of reference, so that all things receive respect. The fruit can be the good things we protect and preserve, and the sweet and tart flavors will be our experience of each other as co-PIE venturers.

GOVERNMENT AS ORGANIZER OF OUR SOCIAL ECOLOGY

It is hard not to feel, surveying the possibilities and problems of using government for the noble purposes of serving everyone, that because things go wrong and fall short so frequently, that this is a foolish if not impossible dream. Hence, it seems that to many people it is best just to keep government to a minimum and rely on people, including companies, to lead the way.

But this is foolish in the extreme for so many reasons. One is that there is no reason to think that people in companies will behave in a way that benefits everyone. Their mission is to make money for the organization, and it is that constricted context that causes the problem of the market failing to take care of the whole. Another is that there is no other player that can do what must be done, but government. It is a unique position government has—it can help people to organize themselves. Thinking of it this way reminds us that in a healthy democratic system, based on cooperation and not battle against two sides, government is us.

It is this view of government concerning which there is no harm and much to be gained. It is this role of government that fulfills its place as the organizing actor in what can be seen as a social ecology. We should not look to government to just give us new rules. We should and must look to government to bring us together so that *we* make the rules that make the most sense for everyone, so that *our* governance shapes our economies to provide peacefully and sustainably for all.

While some may hear this formulation of the relationship between citizen and government as a welcome, doable, wonderful thing, others may conjure up the fear of the mob from this vision. This is when we who wish to see progress can call upon government to articulate and ensure respect for all, to prevent the powerless from being left out and the powerful both from their own overreach and from resentment. If enough people come to see and revere and expect this role of government—as mediator, facilitator, organizer of our best and fairest combined intent—it may be possible that the old idea of government belonging to the winner of a contest, who then gets to have their way, all up and down the system, disregarding the needs or desires of the "losers," can be discarded on the trash heap of history where it belongs.

AN ECOLOGICAL ECONOMICS

Juan Martinez-Allier's 1987 *Ecological Economics* notes that there were a number of thinkers since the beginning of the industrial revolution, who got the big picture and strove to wake people up to the need to reorganize our approach to living on earth. Rudolf Clausius, Serhii Podolinski, Edward Sacher, and Patrick Geddes saw biology as the transformation of the energy of the sun, and that it is the foundation of our wealth. Leopold Pfaundler, in 1902, wrote about the carrying capacity of earth. Frederick Soddy argued that wealth is a "flow, which cannot be saved, but only spent." Wealth growing through compound interest, the foundation of financial operations, is a convention, and is bound by physical reality. Commercial activity takes place within an actual biophysical reality that cannot be ignored, and which, in contrast to the context of finance, is finite. The purpose of the American Economic Association is "The encouragement of economic research, especially the historical and statistical study of the actual conditions of industrial life."[33] The purpose of the International Society for Ecological Economists is:

> the advancement of our understanding of the relationships among ecological, social, and economic systems and the application of this understanding to the mutual well-being of nature and people, especially that of the most vulnerable including future generations.[34]

One of the principal figures of ecological economics is Herman Daly, who asks, "Why do economists insist on treating information and creative works as scarce—while making the opposite mistake with respect to the depletable services of nature, which they treat as limitless by pricing at zero?"[35] David Bollier reports on what Daly told a Forum on Society Wealth at UMass Amherst in 2005:

> Our current economic system treats natural resources as essentially limitless, said Daly, but of course, after a sufficient period of use, most fruits of nature (oil, timber, clean water, air) can be used up. It therefore makes sense to impose "a regime of excludability" on certain natural resources before they become scarce. One tool for doing this is a cap and trade system, which allows governments to prevent the tragedy of an open access regime (often confused with the commons). A cap on usage prevents over-exploitation of a resource, while trading allows people to maximize its efficient use.
>
> In the information commons, by contrast, intellectual property law is used to make an essentially limitless resource—knowledge—scarce. The over-propertization of knowledge can have lots of unfortunate effects, from preventing universal access and benefit to inhibiting the development of new knowledge. Economists see the imposition of artificial scarcity on knowledge (via copyright and trademark law) as a necessary condition for enabling market exchange. But the upshot, said Daly, is that "we mistakenly think that scarcity increases public wealth." In fact, its chief result is the creation of private wealth.[36]

---

33. https://www.aeaweb.org/AboutAEA/gen_info.php.
34. http://www.isecoeco.org/about/constitution/.
35. Daly's books include *Steady-State Economics, Beyond Growth* and *For the Common Good.* David Bollier, "Herman Daly on the Commonwealth of Nature and Knowledge," *On the Commons*, October 3, 2005, https://www.onthecommons.org/herman-daly-commonwealth-nature-and-knowledge.
36. https://www.onthecommons.org/herman-daly-commonwealth-nature-and-knowledge.

SHARED INTELLECTUAL AND SOCIAL CAPACITY

So if we are taken by this idea, and wish to craft a different approach to how we generate and share ideas, what form might it take? In 2000, I attended a conference at MIT held by Thinkcycle, a project of MIT students that focused on the creation of open-source, collaborative design of sustainable technology.[37] At the conference was a presenter from the Honeybee Network of India, a vehicle for sharing information among farmers, innovators, policy makers, and others of like mind. Thinkcycle's collaborative work continues in the form of the P2P Wiki, and the Honeybee Network is now in seventy-five countries.[38] Its website explains that

> what has made Honey Bee Network tight knit and efficiently functional is its philosophy. Honey Bee signifies a philosophy of discourse, which is authentic, accountable and fair. The Network has been woven around three basic ideals. The Network believes that a knowledge system in order to become sustainable has to be both just and fair. Hence, while collecting knowledge from the knowledge holder, the Network has made it a norm to acknowledge the knowledge provider with name and reference, if otherwise not desired by the knowledge provider. This particular practice has come handy in protecting the IPR (intellectual property rights) of the knowledge provider. In the second place, the source of knowledge . . . the traditional knowledge holders and grassroots innovators must be acknowledged . . . any proceed that accrues from the value addition of local traditional knowledge and innovation; a fair and reasonable share must go back to the knowledge holders. These have been the guiding principles of the Network, which are fundamental to the functioning of the network and constitute the major non-negotiable for the Network.

I submitted a poster to the conference, which was displayed on the wall as one entered the auditorium. It was a proposal for an Open Suggestion Box and the Honor Society. The Open Suggestion Box was as simple an idea as it sounds—that there should be a place where people can submit ideas. The Honor Society was a little more complicated. It was the idea that we should honor those who come up with good ideas. We should honor them in any way that we can. If, for example, we took a good idea from the Open Suggestion Box and we put it to good use, and something came of it, "proceeds" as the Honey Bee Network calls it, you would send something to the idea originator, to honor them. The Open Suggestion Box would have recorded that contributor's name and contact information. The Honor Society might engage in investigations to try to learn about the origins of ideas, to guard

37. https://wiki.p2pfoundation.net/ThinkCycle_-_Governance.
38. http://honeybee.org/aboutus.php.

against someone putting their name to another's idea when submitting to the Suggestion Box, because the Honor Society would care about honoring the original source.

Were such an organization to exist, either as a formally organized entity or as a virtual entity, in which people took it upon themselves to investigate and to honor, then the generation of good ideas that benefit us all would be incentivized. I envisioned an Honor Roll of verified good idea originators, and in my best dreams I imagined the concept taking off and millions of people adopting the Honor Role and taking it upon themselves to find the original idea generators of good ideas. Unlike the patent and copyright system, where the first to have a good idea is the only one who has rights to what proceeds from it, I envisioned the act of honoring those who originated the idea, those who took good ideas and elaborated on them, those who found ways to make them real, and those who independently came up with the idea, even if not first. This might be difficult to sort out but the principle seems worth shooting for. The point is that it is ideas and their best implementation that will save us. That we want to incentivize generating and using good ideas. We can redesign our system for doing that. We can foster cooperation and sharing because that is what will save us, not exclusive ownership. It is honorable to give, to provide. It is honorable to be grateful.

The system we now have for incentivizing action, for providing purpose in life, is reductive in value. We honor those who make money by letting them have ownership and power over others. We honor those who reach certain positions without regard to the contributions of their underlings. We honor the rich and powerful and the attractive and the famous, when it is quite possible that they deserve it, because they are people, but it is also quite possible that they don't deserve it.

COMMUNAL PURPOSE OUTSHINES ABSURD POLITICAL DIVISION

If one determines not to turn away from the fearful issues, because we are better off if we face them, the way forward is clear: engaging with others to work them out. Environmental issues affect us all. Recognition of our commonality can reinforce the sense of communal purpose.

Political antagonisms become silly next to civilized and purposeful exchange. When sober adults deliberate sensibly, immature behavior—a fair characterization of much current political argument—loses influence. The fearful usually tempted to join with it see another option, and many choose the better. Environmental issues show us that the ground on which we stand is shared. They can be likened to a dying Mother Earth asking us to be good children to her. They speed the arrival of recognition that enables us to begin the work of the new day. Goodwill in a market society—to which we have been reduced in too many ways—is widely distributed, often invisible. It seems irrelevant to the relationships between members in most contexts, where dollars or other factors dominate. The great mass of good intention is quite disunited and much less effective than it can be. But goodwill, coalesced, is powerful.

The putative divisions keep us apart but their relative importance pales into insignificance in the face of problems like global climate change, loss of species, toxics in our bodies, the nuclear threat. Many legislators are now rewarded for fostering disunity but a cultural movement of healing the body politic could bring revelation of their damaging effect on us all. Uniting the widespread intention for good—tapping the reservoirs of goodwill that exist around us—shows these divisions to be meaningless in the face of the critical needs and opportunities before us all. The cultivation of a shared identity transcends and illuminates, dispelling the shadows cast by the cultural warriors.

The environmental impact caused by humans is an issue that essentially and existentially concerns every human. Effectuating a universal consciousness involves changing identities from sub-universal alone to a dual citizenship, in which every human, every member of a group, is also a member of the species and a citizen of the world.

Having seen success, I think it is possible for the societies of the world to grow the civilization humanity needs. The too-widespread assumption that we need benevolent dictatorship, or some technology or business leaders to save us, is childish and foolish in the extreme. The only thing that can save us is us. A strengthened sense of shared existence and the forms of governance and relationship that come from that are what present the possibility of salvation. We are free to craft new systems of rule-making, of common investment in time and common wealth, out of a conjoining of minds dedicated to the most overall good through the fulfillment of each member of the great body of life. When compared to other things which we could be doing, it makes sense. Going for what you really need is a better plan than compromising at the outset.

Index

About the Author

When as a student in the first class at Hampshire College in 1970, **Rick Reibstein** was permitted to create his own course of study, he began a lifelong effort to understand how to avert further damage from modern civilization, ranging from exploring alternative technologies to the underlying concepts and stories that inhibit or inspire progress. He did not find paying work in the field for many years, but continued mostly voluntary efforts until, in 1987, he found "source reduction" and the sense of accomplishment that came with developing it for the public as staff at a government agency.

Reibstein served for more than two decades as a manager of the Massachusetts Office of Technical Assistance and Technology, one of the nation's first pollution prevention programs, which he helped design and establish. In this position, he created and provided training for companies and others throughout the state, and provided training and consultation on how to provide pollution prevention assistance to EPA, other states, and international visitors. EPA's former international program sent him to train environmental agencies in Brazil, Chile, and the Philippines, and the World Environment Center sent him to the Baltics. Reibstein also served as an enforcement attorney for both the US Environmental Protection Agency and the Massachusetts Department of Environmental Protection.

On the side, he created and provided courses on environmental law and pollution prevention, first for Northeastern University's Continuing Education program, then for Clark, Suffolk, and MIT. Since 2000, he has taught at Boston University, where he introduced the following courses: Law for Sustainability, Developing Sustainable Environmental Responsibility, Research for Environmental Agencies and Organizations (see projects at www.bu.edu/rccp) and the most recent Reconstructing Environmental Governance.

Reibstein originated and provided training to real estate professionals throughout New England (and in Omaha, Nebraska) on environmental matters, most particularly explaining why it makes so much sense to make the effort to make residences lead-safe. In his blog *The Environmental Citizen* he writes to inspire public discussion about environmental policy matters that are not too remote, opaque, or complex to understand, hoping to spread the sense that we are all environmental citizens.[1] Sometimes he gets to facilitate "public conversation" on lead poisoning, a plague we need not tolerate, trying to send a message that our will can be persistent in meeting the threat of persistent poison.[2] For the group Quiet Communities, he is helping to further the national Quiet Conversation needed to address excessive noise, an underappreciated health threat (and linked, such as through gas combustion, to other very harmful pollution).[3]

When he has a choice, he reads, cooks, looks forward to seeing his children, takes long walks with his beloved wife, swims with her in Walden Pond, keeps in touch with dear friends, takes pictures of nature and shares them with friends and family, writes poetry and stories that most people say they like, and plays piano that only some people say they like. He is a "lazy gardener" who fertilizes with his own compost, having traded the lawn for fresh access to greens, spices, tomatoes, and lemon cucumbers in season.

He is anxious that the ideas in this book be taken as suggestions for further thought and work, because this review is rooted in subjective experience. He wishes that experience of constructive engagement can be experienced by many others—by enough to turn the tide.

1. www.trunity.com/ec-blog
2. See meeting reports at: https://www.bphc.org/whatwedo/healthy-homes-environment/lead-poisoning-prevention/Pages/Lead-Poisoning-Prevention.aspx.
3. Contributing the idea of a petition to restore the federal Office of Noise Abatement and Control. https://quietcommunities.org/petition-to-the-epa-to-implement-the-noise-control-act/.

www.ingramcontent.com/pod-product-compliance
Ingram Content Group UK Ltd.
Pitfield, Milton Keynes, MK11 3LW, UK
UKHW012118150926
475560UK00002B/162

9 781538 160039